Praise for *The Bird in the Waterfall*

"A passionate appreciation for the magic, music, and poetry of water, and an appeal for the protection of this most precious of the earth's resources."
—*Natural Resources and Wildlife Magazine*

"Jerry Dennis is one of today's most readable and informative nature essayists, and his latest book, *The Bird in the Waterfall,* is a marvelous look at the natural history of oceans, rivers, and lakes. It ought to be required reading for anyone who loves the outdoors, angling, surfing, beachcombing, or birding."
—*Buffalo News*

"*The Bird in the Waterfall* is truly science for everyone. When you have finished reading it, you will not only know more, but you may become as charmed with water as Dennis is."
—*Earth Magazine*

"I can't think of anyone I know—angler, conservationist, scientific reader, curious kid—who wouldn't enjoy, and learn from, this unusual book. And from endpaper to endpaper, it's a visual delight, too."
—*Fly Rod and Reel Magazine*

"A lovely natural history... written with a grand, entertaining style."
—*Detroit News*

"Dennis teams once more with artist Glenn Wolff to create a wonderful compendium of fact and folklore, historical drama, and personal anecdote about earth's most marvelous compound."
—*Michigan Out-of-Doors*

"Jerry Dennis's clear-eyed essays surpass mere explanation of facts; he conveys the rare gift of understanding the workings of nature, along with passion for its beauties and terrors. His prose is admirably paced with Glenn Wolff's artistic microcosms."
    —*Arts Borealis*

"Jerry Dennis knows water. His new book is a tribute to the magic, music, and poetry of water and an appeal for the protection of this most precious of the earth's resources… a perfect choice for adults and kids alike who want to discover more about how the world is put together."
    —*Lake Country Gazette*

"Nature writer Dennis conveys his deep feelings for all aspects of the aquatic realm… and parlays his fascination with the dynamics of bodies of water into a richly informative description of how lakes and rivers support myriad life-forms."
    —*Booklist*

"Like all fine nature writing, this book instills a renewed sense of wonder for our natural world. A simple drink of water will never be the same again after reading *The Bird in the Waterfall*."
    —*Flint Journal*

"A masterful work of natural history [that] exerts a steady and inexorable pull. This read is well worth your while."
    —*Traverse Magazine*

The

BIRD
in the
WATERFALL

Exploring the World of Water

The
BIRD
in the
WATERFALL

Exploring the World of Water

JERRY DENNIS

drawings by
GLENN WOLFF

THE BIRD IN THE WATERFALL

Revised paperback edition published 2014.

Originally published by HarperCollins Publishers, 1996.

Copyright © 1996 by Jerry Dennis

Illustrations Copyright © 1996 by Glenn Wolff

ISBN: 978-1-940941-52-3

To Craig Date, waterfall jumper
—J.D.

To my family, creatures of water, every one
—G.W.

CONTENTS

INTRODUCTION   1

1 THE CYCLE OF WATERS   3

2 WATER AND LIFE   13

    Evaporation and Transpiration   20

3 WATER UNDERGROUND   25

    Dowsing For Underground Water   33

    Aquatic Life Underground   35

    The Deepest Groundwater   36

4 THE MAGIC OF SPRINGS   39

    Life in Springs   45

5 IN HOT WATER: THERMAL SPRINGS AND GEYSERS   49

    Taking the Waters   53

    Life in Hot Water   58

6 THE BIRTH OF A RIVER: BROOKS AND CREEKS   61

    Life in Small Streams   63

7 MEANDERING RIVERS   69

    The Science of Meandering   75

8 THE ANATOMY OF RIVERS   79

    The Velocity of a River   83

    Reading a River   85

    Stream Capture   88

    Ephemeral and Intermittent Streams   90

    Life Beneath a River   93

9 THE BIRD IN THE WATERFALL   95

    Life in the Fast Lane   103

10 OVER THE WATERFALL   107

11 THE COLOR OF WATER   119

    Colorful Names   122

12 ICE   127

    Why the Ice Booms   132

    Other Forms of Ice   135

    Sea Ice   137

    Glaciers, Ice Sheets, and Icebergs   139

13 SKY WATER: PONDS AND LAKES   143
     Very Large, Very Old Lakes   149
     Shallow Lakes   154
     Wetlands   156
     Life in a Frozen Lake   159
     Seiches   161
14 LIFE ON THE SURFACE   165
     The Amazing Whirligig   171
15 SALT LAKES AND SEA MONKEYS   173
     The Holiest Salt Lake   181
     The Death of a Salt Lake   182
16 MOTHER OCEAN   185
     Sea Life   194
     Rivers in the Sea: Ocean Currents   199
     The Abyss   203
     Sound Underwater   206
17 SALT WATER   211
18 WAVES   219
     Freak Waves And Rogues   226
     Oil On The Water   230
     Storm Surge   231
     Waves Underwater   232
     Tsunamis   234
19 THE TUG OF THE MOON: OCEAN TIDES   241
     Tidal Bores   247
     Tidal Races, Whirlpools, and Overfalls   248
20 THE DYNAMIC BEACH   251
     Patterns on a Beach   256
     Rip Currents   259
Selected Bibliography   263
Index   267

INTRODUCTION

Many years ago, in an alpine meadow high in the Montana Rockies, I kneeled beside a creek to take a drink. The creek was fed by snowpack in a pass above me and flowed with water so clear that every pebble on the bottom showed with the kind of magnified clarity you expect to see only with fine optical equipment. I leaned over to drink and was startled when my face plunged into icy water. The water was transparent as air, hardly more than a disorganized shimmering of light and shadow, and I could not see where atmosphere ended and water began until I crashed the boundary. It was the clearest stream I had ever seen, and the sweetest tasting. I drank my fill, then lifted water by the handful and watched it run through my fingers. It was like coming face to face with the origin of the word *purity*.

This book was written with that stream in mind.

Earth is a water planet—about 70 percent covered with it. We inhabit islands of rock and soil surrounded in all directions by water. It permeates the ground, the atmosphere, and every living thing. We humans are mobile reservoirs of it, carrying it everywhere we go. We never seem to get enough of it. When we vacation we travel almost automatically to oceans, lakes, and rivers. For recreation we swim, fish, raft, canoe, sail, surf, ski, or scuba dive. We relax by taking hot baths or showers. When we need time to reflect and replenish, we prefer to go to places where we can watch waves or current or falls.

Yet, in spite of water's central role in our lives, we have treated it poorly. For most of the history of civilization we have dumped our wastes into the nearest water, leaving our fouled trails behind. In these relatively enlightened times we like to think we're too smart to urinate in the well, but the evidence doesn't support it. We still contaminate with waste, still deplete rivers and aquifers to over-irrigate our crops, still flush fertilizers and pesticides into every waterway. As of 2014, more than a billion people lack access to safe drinking water, and every day six thousand children die of waterborne illnesses. It's clear that our water resources will continue to be tested. And it's increasingly crucial to remember how fragile those resources are and how much they need to be protected and conserved.

For about half of every year I can step outside my house and see water in its solid, liquid, and gaseous forms. That it can be found in all three forms at once so impressed the Greek philosopher Thales that he decided water was the fundamental element of the universe, more essential than fire, air, or earth. All matter, he argued, must be composed of water because anything that was not made of it was born in it or changed by it.

We sense something of Thales's esteem when we see in a single day— sometimes in a single glance—a stream, a field of snow, a frozen lake, dripping icicles, a sky obscured with clouds. If we live in a place where water is abundant, it's possible to see it in all its forms and begin to take it for granted. But we do so at our peril.

It's my hope that this book will serve as a reminder that water is a wonder—the most wonderful substance on earth—and that it is absolutely essential to the survival of all living things.

1

THE CYCLE OF WATERS

I was hiking across snow-covered meadows in January when I found a stretch of East Creek I had never seen in winter. It was small, three or four feet wide, and made even smaller where snow had drifted beyond the banks and built canopies over the water. The effect was startling, the structure so fragile I hesitated to approach. Water flowed somewhere beneath the snow—I could hear it, like a stream in a grotto—but I had to lean out carefully from the bank before I could see that the creek was clear and quick, with a bottom of bright gravel.

Near the water, a line of tiny icicles clung to the snow like drops frozen to the beard of an arctic explorer. As I watched, a drop formed, paused for a moment on its tip of ice, then fell into the creek. I imagined the molecules within that drop unfurling and drifting downstream and joining the Boardman River, then flowing through riffles and pools and impoundments to West Bay. From the bay they would drift north up Lake Michigan to the Straits of Mackinac, pass beneath the bridge, then make slow progress south the length of Lake Huron. They would enter the St. Clair River and pass through Lake St. Clair to the Detroit River, then cross shallow Lake Erie to the Niagara River. They would become airborne at

Niagara Falls, settle as mist in the rapids below, and tumble downstream to Lake Ontario. They would drift eastward to the outlet of the lake, downstream past the Thousand Islands, and into the broad St. Lawrence, passing Ogdensburg and Trois-Riviéres and Québec City until finally they merged with salt where all river journeys end. I focused again on the creek, bewitched by water, pleased with the tidiness of its complete world. Then something cool and wet touched my cheek and I looked up: snowflakes.

It is easy to forget that snowflakes and creeks are links in a hydrologic system that circulates water constantly between the earth and the sky. It is a system without beginning or end, a true cycle, and it is complex, covering many possible paths. The system is so extensive that it is proper to label its domain the *hydrosphere*, a watery realm that extends from a dozen miles above the ground to a few thousand feet beneath it and includes all the water in rivulets and rivers, in icicles and polar ice fields, in clouds and desert springs and oceans.

At any moment, a little more than 97 percent of the 326 million cubic miles of water in the hydrosphere is contained in the oceans. The remaining 3 percent is fresh water, about three-quarters of which is locked up in glaciers and ice sheets at the poles. Most of the rest of the freshwater supply is in the ground. Only about 0.036 percent of the earth's water is contained in lakes and rivers, and a much smaller amount—about 0.001 percent—is in the form of vapor in the atmosphere. If all the water vapor were to condense at once and fall as rain it would flood the surface of the planet with about one inch of water. In contrast, if the earth were a smooth sphere and all the water in the oceans could be distributed evenly across its surface, it would be covered to a depth of 8,800 feet.

But of course water is not distributed evenly. Abundant in some places, scarce in others, it moves constantly as solid, liquid, and gas through countless channels within the hydrologic cycle. Every day 210 cubic miles of water evaporates into the atmosphere from the oceans, and another 38 cubic miles enters the sky as evaporation from land and inland waters and as transpiration from plants. An equal amount condenses and falls back to earth as precipitation. Most of each day's evaporation and precipitation takes place over the oceans in a simple exchange between sea and sky. When precipitation falls on land or ice, however, there are many possibilities for sidetracks and delays and the cycle becomes more complicated. A molecule of water might drift to the ground in a snow crystal and become enveloped in a glacier that take hundreds of years to creep through a mountain pass to the ocean. It might descend as rain that soaks into the ground and is picked up by the roots of a tree, is carried to leaves high in the crown, and transpires as vapor back into the air. It might fall during a desert rainstorm and run off into a roaring, clay-colored arroyo that for 360 days each year is a gash of dry sand. It might sink into soil and enter an aquifer. It might be swallowed by animals or piped into septic tanks or treatment plants. It might get sealed inside a plastic bottle and placed on a supermarket shelf. Always, regardless of the paths it follows or how harshly it is used, water completes the cycle from ocean to sky to earth and back to ocean.

The period of the cycle varies greatly. On average, a molecule of water spends only nine days as vapor in the atmosphere before condensing as frost or dew or falling as rain, snow, sleet, or hail. When it finds its way into a river (a typical one, with a current speed of about three feet per second) it stays there about two weeks. If it soaks into the first few feet of the ground it remains for as little as two weeks or as long as a year before it evaporates or is absorbed by plants. If it infiltrates deeper and enters the shallow groundwater supply, it will stay there tens or hundreds of years. If it infiltrates farther yet and enters the deep groundwater it won't come to the surface again for thousands of years. It is likely to stay in a large lake for 10 years, the shallows of an ocean for 120 years, and ocean depths for 3,000 years. If it falls on the Antarctic ice cap it will remain there perhaps 10,000 years before riding into the ocean.

Through all its routes, the hydrologic cycle maintains a broad equilibrium. As fast as water runs into the seas, it returns to the sky to journey through the cycle again. Each year the oceans lose enough water through evaporation to reduce their levels by more than four feet. If the oceans were not replenished, in 4,000 years they would be dry, salt-encrusted chasms. But the lost water is always replaced by rain and snow (enough to return about 3.7 feet to the oceans) and the discharge of rivers (enough to return about 0.3 feet). Likewise, though the land surfaces of earth receive enough rain and other precipitation each year to cover the continents with 29 inches of water, the continents are not constantly flooded because about 17 of those inches return quickly to the atmosphere through evaporation and transpiration and the remaining 12 inches infiltrate the ground or are carried away as runoff into rivers.

The problem with such figures is they make the hydrologic system appear more orderly than it is. Local and seasonal variations in rainfall can cause floods or droughts so destructive they make the concept of aquatic equilibrium ludicrous. As climate changes over millennia, the distribution of water changes with it, transforming a lush forest into the Sahara, shrinking an inland sea the size of Lake Michigan into Great Salt Lake. But when water no longer falls on the Sahara, it is because it is falling more heavily elsewhere. All the amounts average out because, strictly speaking, there is no way to take water away or add water to the cycle. Virtually the entire supply circulating through the ground, on the surface, and in the air has been here since the infancy of the earth.

Wherever we intercept the water cycle, it seems complete. Rain appears to be an isolated phenomenon, the oceans are full and permanent, a snowbound creek is a settled feature of the landscape. It takes a broad view to see the whole system at work.

That broad view is a relatively recent accomplishment. The Greeks and others in ancient times wondered much about the origin of springs and rivers but only flirted with the idea of a water cycle. Those who debated the source of rivers usually divided into two camps. Some took the side of Plato, who argued that water from the ocean was carried to the Underworld by the River Styx, where it was filtered of its salt and divided into small streams that returned to the surface as springs. Others sided with Aristotle, who proposed that fresh water was formed when air and earth combined in caverns deep underground, creating new water that later found its way to the surface.

Later, the Roman philosopher Seneca compared the waters that circulate through the oceans and cavities beneath the land to the closed system of blood that circulates through the veins, arteries, and cavities of the human body. Yet neither Seneca nor most other observers in the Mediterranean region recognized that rain and snow were the source of springs and rivers. Rain was intermittent, after all, and fell in quantities that seemed insufficient to feed mighty rivers like the Nile or the Rhône.

About 100 B.C., the Roman engineer Marcus Vitruvius Pollio proposed a cyclical nature for water that was unique in giving precipitation a role. In a section of his *De architectura* describing techniques for drilling wells, he stated matter-of-factly that the source of groundwater was rain and melted snow that had infiltrated the earth. He also explained that water on the surface is changed to vapor by the sun and rises into the sky to become clouds, which eventually collide with mountains and break open, causing rain to spill out.

Such a revolutionary notion was slow to be accepted. For centuries the popular imagination seems to have been stirred most by Plato's model of water entering the earth through caverns beneath the oceans, where it was filtered of salts before finding its way by mysterious processes to the surface. In the Middle Ages it was thought that the oceans did not overflow because all their excess water drained away into a reservoir at the center of the earth, which became the source of springs, rivers, ocean currents, and tides. Water gushed to the surface the way blood gushed from an open wound, defying gravity because it was powered by heat within the earth, or by magic, or by some undiscovered law of physics. By the seventeenth century, the popular concept of the "living earth" was used to argue that water in the ground was a byproduct of the planet's respiration. To explain the baffling but easily observed fact that groundwater existed far above the surface of the sea, some scholars proposed that the curvature of the earth actually placed the ocean's center of gravity far above the land, so that the emergence of springs at even the highest elevations was caused by nothing more complicated than water seeking its natural level. Even clues in the Bible and other sacred texts were overlooked. The observation in Ecclesiastes that "all the rivers run into the sea; yet the sea is not full; unto the place from whence the rivers come, thither they return again," has rarely been interpreted as natural history.

In an era when most people in western Europe believed that surface water originated from the oceans or was manufactured magically in the center of the planet, Leonardo da Vinci (1452-1519) proposed that clouds, "the begetters of rivers," were part of a worldwide cycle of waters. In his notebooks he wrote:

> You can well imagine that all the time that Tigris and Euphrates
> have flowed from the summits of the mountains of Armenia, it must
> be believed that all the water of the ocean has passed many times
> through these mouths. And do you not believe that the Nile must have
> sent more water into the sea than at present exists of all the element of
> water?
>
> Undoubtedly, yes. And if all this water had fallen away from this
> body of the earth, this terrestrial machine would long since have been
> without water. Whence we may conclude that the water goes from
> the rivers to the sea, and from the sea to the rivers, thus constantly
> circulating and returning, and that all the sea and the rivers have
> passed through the mouth of the Nile an infinite number of times.

Like many of Leonardo's insights, this one was so contrary to popular opinion that it made no impact during his lifetime. Not until the middle of the seventeenth century were the first scientific tests of the hydrologic cycle as we know it undertaken. Two French scientists, Pierre Perrault (1608-1680) and Edmé Mariotte (1620-1684), working independently of one another, measured the precipitation that fell within the drainage basin of the Seine River and found that it was more than enough to account for the amount the river discharged into the English Channel. About the same time, British scientist Edmund Halley (1656-1742), who is best known for discovering the comet that bears his name, calculated the amount of water that evaporated from the Mediterranean and concluded that it was approximately equal to the amount that flowed into the sea from its tributaries. Unlike many of the scientific discoveries of the era, those early models of the hydrologic cycle were given approval by the Church because a self-contained system of water could be explained by the Christian doctrine of the Great Divine Order, in which all of nature was created for the use and delight of mankind. What, it was reasoned, could be more delightful and useful than a closed system of constantly renewed water?

People intuited a cycle of water and gave expression to it in their mythologies and legends long before scientists tested, described, and named it. To the early Greeks, the earth was surrounded by Oceanus, a vast and endlessly flowing river that was represented circling the land like a snake with its tail in its mouth. Classical Eastern cultures portrayed most of their sacred waters flowing in circles. The ancient Egyptians saw a river running through the heavens, with the sun and the moon riding the current around the earth every day and night. In many mythologies rivers issue from the earth, the mother who gives birth to generation after generation of living things. The cycle of waters is paralleled with cycles of birth and death, with the cycle of the seasons, the sun and moon, and day and night. All things come and go, then come and go again, and the waters always run.

In the folklore of the Great Lakes region Paul Bunyan organizes a log drive on Round River, a waterway that flows into itself and circles in a continuous loop somewhere in the woods of northern Michigan or Wisconsin. In most versions of the legend, Paul Bunyan leads a crew of lumbermen who cut and stockpile a winter's worth of logs on the banks of the river, then in the spring ride the logs downstream toward Lake Michigan, trusting that a sawmill will be found along the way. After two weeks Bunyan and his companions float past an abandoned camp that looks suspiciously like the one in which they spent the winter. Two weeks later, when they pass the camp again, they realize their log drive can never end.

Such circularity speaks partly of the seasonal aspects of lumbering, a job that must have seemed endless to lumbermen who cut and drove logs year after year in forests most people thought could never be depleted. But the legend also gives voice to the self-contained and cyclical world of nature. To conservationist Aldo Leopold the tale of Round River was a parable for the biosphere—the flow of energy from soil to plant to animal and back to soil—and an argument for a sound conservation ethic. In Leopold's view we are circling in a perpetual stream of land, water, and life, and every part of it is critical to the fertility and health of the whole. Central to that flow of energy and life is the hydrologic cycle.

In "The Round River Drive," a rhymed version of the Paul Bunyan tale written in 1914 by a poet named Douglas Malloch, Round River is located in "section 37" (a fallacious site, since townships consist of 36 sections) in the woods of northern Michigan, "...west of Graylin' 50 miles." It happens that I've lived most of my life in northern Michigan, almost precisely 50 miles west of Grayling. Many of my favorite streams are located there, including the small one I came across that January day when the banks were dripping with icicles. I've explored the region in all seasons and have yet to come across a river that flows in a circle, without source or conclusion. If that is a failure, it is a failure of imagination, not discovery. In the larger context there are no beginnings or endings. Snowflakes fall into creeks that flow into oceans that rise to the sky. Every drop of rain carries a bit of sea. Every river is a round river.

2

WATER AND LIFE

We're fascinated with water. We can't help ourselves. It should not be a surprise, given the intimate relationship between humans and water—we're 70 percent made of it, after all—but our relationship with food is equally intimate yet you seldom see people lining up to stare at fields of corn. There is something special about water. It is common but not commonplace, the most abundant compound in the upper crust of the earth, six times more abundant than feldspar, the most abundant mineral. It seems so simple: fluid at normal temperatures, frozen when cold, vaporous when heated— too transparent to be complicated and too plentiful to be precious. But appearances deceive. Water is among the most remarkable of all compounds.

Science and mythology agree that in the beginning there was water. Four billion years ago, when the earth was a mere half-billion years old, the atmosphere was filled with clouds of water vapor, hydrogen, ammonia, methane, and other gases, all drawn by gravity to the young, cooling planet. Water was released as steam from the molten rock beneath its surface and brought to earth as ice on comets that melted as they plunged through the sky, spraying water across the new planet. In the atmosphere, water vapor rose until it cooled, then condensed into droplets and fell as rain. As it fell, it dissolved atmospheric gases, carrying to the oceans the ingredients of the "primordial soup" so central to many scientific discussions of the origins of life. It rained volumes unimaginable to us. It rained for thousands of years. It rained until high volcanic mountains had cooled and the lowlands between them had filled to become oceans.

Much of the credit for the continued existence of those oceans must go to luck. It was our good luck that in a universe that varies some 36 million degrees in temperature from the unimaginable cold of deep space to the equally unimaginable heat of stars, water found an environment on earth where it could exist in a fluid state in the mere 180 degrees Fahrenheit between freezing and boiling. The luck held and the oceans filled. In time, the conditions were right for life to appear.

This notion of a young planet awash in life-giving water carries uncanny echoes of creation myths from all over the world. In a Babylonian myth the universe was a single enormous ocean and the earth a mat of rushes woven by the hero-god Marduk, who heaped soil on it and set it afloat. In the *Iliad*, Homer wrote that Oceanus was "the source of the gods," and "the source of all things." In Hebrew and Christian traditions God divided light from darkness, creating heaven and earth and lifting them from the chaos of infinite waters. The ancient Egyptians believed the earth was created from the abyss of formless waters known as Nun (or Nu), and all land floated on the remnants of that sea. Finnish, Japanese, Iranian, and North American Iroquois and Cherokee myths all speak of a universe of infinite water, on which the earth floats like an island or was formed when gods or animals plunged to the bottom of the primordial waters and emerged clasping soil in their hands.

Scientists are nearly unanimous in believing that life first appeared on earth about three and a half or four billion years ago and that it began in water. For more than a century Charles Darwin's "warm little pond" with its broth of ammonia, formaldehyde, cyanide, methane, hydrogen sulfide, and other compounds was considered the most likely site for the first life. Recent theories have challenged Darwin's notion, proposing that life may have first appeared near hot vents deep in the oceans or in bubbles of foam or films of oily molecules on the surface. Still, most theorists agree that the first life probably appeared in liquid water because only there could molecules move freely and randomly enough to combine in ways that made life possible.

The origin of life will likely remain uncertain and controversial, but how it is sustained is clear. Water's ability to dissolve most compounds allows it to maintain life in plants and animals by transporting nutrients to their cells and carrying away their wastes. In animals, transportation of nutrients, oxygen, and hormones is through a circulatory system composed mostly of water. In plants, water carries nutrients and oxygen from roots to leaves and is a primary component in photosynthesis, in which sunlight converts water and carbon dioxide into carbohydrates and oxygen. Photosynthesis breaks water molecules apart, the hydrogen becoming part of the carbohydrate compounds, the oxygen freed to enter the atmosphere. Without water there could be no photosynthesis, and without photosynthesis there would not be enough oxygen in the atmosphere to support animal life.

At birth we are 90 percent water. We dry as we age until, by maturity, our bodies are about 70 percent water and our blood 83 percent, for a total volume of about 10 gallons. It flows through our veins, lubricates our organs, fills our cells and the spaces between them. We can live a few weeks without food, but only a few days without water. Other than the air we breathe it is our most urgent need. Even breathing and eating require it. We can absorb oxygen into our blood only when it is first dissolved in the moisture in our lungs. When we eat, our food must be dissolved to liquid in our digestive systems before it can be absorbed into our bodies. Much of our food, therefore, is high in water content: A potato contains 80 percent, an apple 85 percent, tomatoes and lettuce 95 percent, a watermelon 97 percent.

Under ordinary conditions we need to replenish our bodies with about two quarts of water per day, but ordinary conditions vary greatly from person to person and place to place. A farm-worker laboring in fields in the American Southwest might need a gallon in eight hours, while an office worker in an air-conditioned cubicle in Chicago can get by with a pint or two. Our overall consumption, however, far exceeds what we need for drinking. On average every person in the United States uses about 1,400 gallons of water per day. About 100 gallons of that is for personal and household use, and the rest goes to irrigate crops and manufacture goods. Elsewhere in the world, water is portioned out much more carefully. In most of Europe, the average amount of water used per person is about one-third what it is in the United States. In developing nations it is not unusual for the average to be as low as 12 gallons per person per day. Worldwide, nearly 70 percent of all the water used goes to irrigate crops—a good thing when increased food production feeds hungry people, but potentially disastrous when too-heavy irrigation results in contamination of the soil with salt, pollution of surface and ground waters, depletion of underground water supplies, and subsidence of surface land.

Our need for water has increased dramatically as populations and industries have grown. In the year 1900, when the world's population stood at 1.5 billion, humans used about 400 billion cubic meters of water, or about 242 cubic meters per person. By 2012 population had increased less than five-fold, to 7 billion, but the amount of water appropriated for human use had increased more than ten-fold.

Water is so critical to us that it has always been one of the primary prerequisites in the development of civilizations. From agricultural settlements along the Euphrates and Tigris Rivers 6,000 years ago to the great modern cities of Europe, Asia, and North America, water has determined the fates of civilizations. Where it is abundant, they tend to thrive; where it is scarce, they tend to fail.

*

Most of the qualities that make water so useful and so central to life on earth can be traced to its molecular structure. Any schoolchild can tell you that water is H20, a simple molecule composed of two atoms of hydrogen attached to a single atom of oxygen. Scientists and kids alike delight in describing a Mickey Mouse molecule with an oxygen head and hydrogen ears. But that pair of hydrogen ears make the molecule lopsided, creating a positive electrical charge on the hydrogen side and a negative charge on the oxygen side. This dipolar arrangement makes molecules attract one another, each hydrogen atom clinging firmly to its own oxygen atom but also clinging more tentatively to the oxygen atom of a neighboring molecule.

It seems a mere quirk, but the hydrogen bond between molecules is the secret to water's remarkable qualities. The bond creates cohesion within water, which is why there is surface tension and why stones skip, pins float, and water striders walk on it. Cohesion also explains why drops form, why rain beads and flows in runnels down a window or a windshield, why ice floats, and why a body of water remains a body rather than instantly evaporating. The attraction of its molecules makes water stick together so stubbornly that when it passes through a hose or pipe it can flow uphill as a siphon. This apparent contradiction to the law of gravity works as long as no air breaks the connection of molecules in the liquid, and makes it possible to steal gasoline from an automobile and move water over hills in aqueduct systems.

When a water molecule forms a hydrogen bond with the molecule of a substance other than water, the attraction is called adhesion. Adhesion explains *capillary action*, the force that causes water to be sucked up by a paper towel and rise in defiance of gravity in a narrow glass tube. In each of those examples, water "climbs" the walls of an adhesive substance, filling the spaces the substance surrounds. Place small-diameter glass tubes in a pan of water, and the water rises in the tubes higher than the surface in the pan. How high depends on the diameter of the tubes. Water in a tube one millimeter across rises about one foot; with a diameter one-tenth smaller it rises 10 feet. If an object gets wet when dipped in water you can be sure it has a surface electrical charge and thus is adhesive to water. On the other hand, waxed paper does not get wet because wax is composed of non-polar molecules that have no attraction to water's molecules.

The polarity of water's molecules makes it the most versatile of all solvents, giving it the power to force the molecules of many other substances apart by separating their positive and negative ions. Water is such an effective solvent that it is virtually impossible to find pure water. Raindrops swirl with dust and dissolved gases and minerals. When they come to earth they acquire a little of everything they touch: dust, leaves, iron ore, stone.

Even the cleanest, most transparent springwater contains dissolved rock and suspended organic matter. Hard water contains 100 or more parts per million of minerals, and probably tastes better than soft water, though it does not allow soap to lather. Soft water has been treated to remove minerals and is latherable but may be unpleasant to the taste. The only pure water is distilled water, made by bringing water to a boil, condensing its steam, and collecting the condensation in sterile containers. Perhaps because we are accustomed to the flavor of minerals in water, distilled water tastes terrible, like stale breath and old cardboard.

Another of water's remarkable qualities is its ability to absorb and store heat. It has this ability because of its *specific heat*, which is the amount of heat that must be absorbed or lost to change the temperature of one gram of a substance by one degree Celsius. Water's specific heat is unusually high, allowing it to lose or absorb tremendous amounts of heat without much change in its temperature. Again, this quality can be traced to hydrogen bonding. Because heat is absorbed when hydrogen bonds break and released when they form, much of the heat energy that enters water is used, not to heat the water, but to disrupt hydrogen bonds. As a result, the oceans are an immense climate-control system. They absorb heat during the day and during hot seasons, then release it gradually during the night and during cooler seasons, serving as a moderating influence on weather. Coastal regions typically experience more moderate temperature than regions inland, but the planet as a whole benefits as well. Without the heat-distributing influence of the oceans, temperatures on the surface of the earth would be dangerously hot during the day and brutally cold at night.

Hydrogen bonds between molecules of water are fleeting at best. Each molecule connects and disconnects with its neighbors 10 billion to 100 billion times every second, forming and dissolving partnerships at incomprehensible speed, and making water highly unstable and in many ways unpredictable. What is known about water could fill a library, yet not even the most knowledgeable hydrologist can predict the pattern of waves in mid-ocean or the path a new stream will follow downhill. Old assumptions about water are often disproved when examined with new techniques and

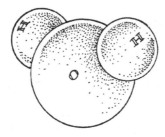

technology. One example: a drop of water contains trillions of molecules of water, all of them drawing together to form a sphere, the most compact shape possible. Scientists have long assumed that the tendency of water to form drops would not apply when the number of molecules was reduced to dozens or hundreds instead of trillions. But recent research with computer simulation suggests that when as few as 90 molecules of water gather they automatically shape themselves into minute drops identical to the much larger drops of ordinary experience.

*

For thousands of years, people have considered this amazing substance magical and holy. Water can so effectively wash the dirt from our hands that it is not unreasonable to think that it can wash the sins from our souls. Consecrated water has purified practitioners of many religions all over the world. Rituals of ablution date to ancient Greece, especially the city of Eleusis, where sacred "mysteries" were performed on young initiates. The mysteries involved a quick emersion in the sea, from which initiates emerged dripping but renewed in spirit and in name. In ancient Japan, the newborn children of the emperors were given holy "Rites of the First Bath," a ritual similar to baptism by some Christians, who believe that a soul does not gain entry to heaven if the body is not first cleansed with holy water. Aztec newborns were cleansed by midwives, who would chant, according to Alev Lytle Croutier in *Taking the Waters: Spirit, Art, Sensuality*, "May this water purify and quieten thy heart; may it wash all that is evil."

The Hebrew claim that cleanliness is next to godliness has been taken to heart by Catholics, Orthodox Jews, Muslims, and many others of the religious devout, all of whom cleanse themselves spiritually by dipping their hands or fingers in holy water before praying to their gods.

In Christian traditions the holiness of water is undisputed, though the question of whether baptism should involve a few sprinkled drops or total immersion has been the source of much dispute. Roman Catholic priests have sanction to create holy water by blessing it, turning ordinary tap water into Easter water or Gregorian water with the power to wash away evil, purify uncleanliness, and eliminate temptation. In ancient Egypt, holy water from the Nile was so essential to sacred rites that the river's water was bottled and carried to Rome to ensure that the rites in the temple of Isis were performed correctly.

We tend to congratulate ourselves for being free of paganism, but next time you're near a public fountain take a few minutes to notice how many people toss coins in the water as they pass. It's an ancient practice. Making small sacrifices to propitiate water spirits would have been one way to ensure sufficient water for the future and to perhaps buy protection against the dangers of too much water. When we drop a penny or a dime in a fountain we make a similar small sacrifice, "for luck."

Even without sacred or magical powers, water remains a wonder. It is a scientific oddity, enigmatic and anomalous and the subject of thousands of years' worth of speculation, yearning, fear, abuse, and veneration. Though literally more common than dirt, it can become so scarce that we value it more than almost anything. It has the power to annihilate cities and inundate nations, yet a few drops of it can urge life to spring from desert sands and sidewalk cracks. It can be enchanting or terrifying, life-giving or deadly. Without it we simply could not exist. It's no wonder that the anthropologist and essayist Loren Eiseley could watch wind ripple across a puddle on a city roof and write, "If there is magic on this planet, it is contained in water."

EVAPORATION AND TRANSPIRATION

Water's ability to evaporate, or change from liquid to gas, has an enormous role in the hydrologic cycle. Evaporation occurs when molecules of water become activated with enough kinetic energy to kick themselves free of the other molecules in a mass of water. That kinetic energy is produced from heat. When water is warmed its molecules move more rapidly and more of them escape into the air, which is why a puddle dries so quickly on a hot day and so slowly on a cool one. Theoretically, there will be no evaporation when the temperature of the water is the same as the temperature of the air above it. When the relative humidity of the air reaches 100 percent, the number of molecules that escape from a body of water are matched by the number of molecules that fall back into the water and there is no net loss to evaporation.

The rate of evaporation varies a great deal according to temperature of the water and air, humidity of the air, the wind, barometric pressure, precipitation, the quality and composition of the water, and the dimensions of its surface. The largest factor contributing to how quickly or slowly a body of water evaporates is sunlight. Solar heat activates molecules near the surface of the water, causing the most active of them to eject into the air. If there is wind, the broken surface of the water and the liberating force of the wind help the molecules escape and speed the process.

As water evaporates, heat energy is converted to kinetic energy, causing a general cooling of the liquid. The faster it evaporates the faster the temperature falls, which is why perspiration feels cool on our skin and why blowing across the top of a hot beverage cools it. In the same way, a hot wind on a summer day will dry a puddle so quickly you can virtually watch it recede.

Salt water does not evaporate as quickly as freshwater. For every 1 percent of salinity, evaporation slows about 1 percent. Seawater containing 3.5 percent salt, then, evaporates about 3 percent slower than freshwater. On the other hand, turbid water absorbs more solar energy than clear water, activating more molecules with kinetic energy and increasing the rate of evaporation.

Evaporation is a crucial step in *transpiration*, the process by which water is transported through living plants to the atmosphere. It is so much a part of the process that scientists sometimes join the two processes together into the single term *evapotranspiration*. Water is essential for photosynthesis and other metabolic activities in plants, but only a small percentage of the water contained in each plant is used for such purposes. More than 90 percent passes straight through from roots to leaves and exits into the atmosphere. The water escapes through tiny pores in the leaves called stomata (from the Greek *stoma*, "mouth"). When the stomata open to allow carbon dioxide to enter, a small amount of water in the leaf evaporates. As the water escapes, it creates a suction force, causing the water tension in the leaf to increase, pulling water from veinlets in the leaf to replace the evaporated water. The water pulled through the veinlets is in turn replaced by water pulled through the woody tissue of the twig holding the leaf. The twig pulls water from a branch, which pulls it from the trunk, which pulls it from the roots. Tiny root hairs absorb water directly from the soil.

Plants are living fountains, sucking water from the earth and spouting it as vapor into the sky. The total amounts spouted are astounding. A large oak tree can transpire 40,000 gallons in a year, an acre of corn 3,000 to 4,000 gallons every day. According to some calculations, more water flows through plants during any period of time than flows through all the rivers of the world. And because water spends much less time in plants than it does in rivers, a significant percentage of the world's fresh water is being constantly pumped from the ground into the atmosphere through roots, stalks, stems, needles, and leaves.

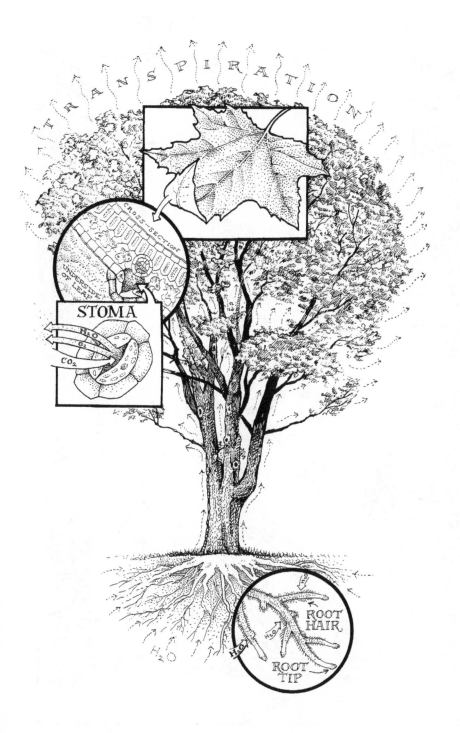

3
WATER UNDERGROUND

There are freshwater seas beneath us. They can't be sailed on and they can rarely be seen, but those subterranean seas contain as much as 30 to 100 times more water than all the rivers and lakes on all the continents. Only the ice caps and glaciers impound a greater supply of freshwater. Dig deep enough, almost anywhere, and you will strike water.

By some estimates more than two million cubic miles of water is stored beneath the surface of the earth. If that entire supply could be pumped from the ground it would cover all land on the planet to a depth of a thousand feet. Some of the water in underground reservoirs is tapped with wells and brought to the surface. Some of it comes up naturally in springs or geysers. Much of it remains untapped and unreachable.

How did those reservoirs fill? A very small percentage of groundwater begins as steam rising from molten rock deep in the earth or became trapped eons ago beneath sedimentary rocks and carried slowly downward. But most of it, virtually all of it, begins as rain or snow that soaks into the ground and "recharges" the groundwater supply.

Soil is a sponge for water. Even when the surface is soaked, dry soil at deeper levels continues to suck water downward. Some water goes no deeper than a few feet, where it is absorbed by the roots of plants. But most of it passes beyond the root zone, draining slowly downward through the pores between grains of soil until it is stopped by a layer of bedrock as watertight at the bottom of a concrete swimming pool. There it accumulates, filling the spaces between stones and sand, and becomes an *aquifer*.

An aquifer is simply a subterranean reservoir of water, but it is inaccurate to think of it as a reservoir in the usual sense. It is, instead, a zone of soil saturated with water, similar to the seeping sand you can find by digging in beach sand near the water's edge. It can be contained in layers of sand, gravel, or porous rocks like limestone and sandstone, at depths of a few feet or a few thousand feet. Its water-holding capacity depends on its composition. Limestone can hold only about half a gallon of water for every cubic foot, while a cubic foot of coarse sand can absorb up to three gallons.

The top of an aquifer is the *water table*. Water on the surface always seeks a uniform level, but the depth of the water table varies from place to place because groundwater is affected by forces other than gravity. In a static world aquifers would rest in horizontal layers. But our world bucks and gyrates, forcing layers of rock to tilt and buckle and fold. An aquifer tends to parallel the contours of the surface, rising beneath hills and sloping down into valleys, but it is subject to vagaries. It might flow downhill for a short distance along the slope of a valley only to be blocked by beds of clay and layers of dense rock. It might become stacked above another aquifer, separated by a stratum of impermeable rock that isolates it as a *perched aquifer*. It might remain trapped until penetrated by a well or until an eroding gully slices down far enough to expose it, freeing its water to leak into the world.

Aquifers can be either confined or unconfined. A confined aquifer is sandwiched between layers of impermeable rock and can be recharged only if it connects with other aquifers or outcrops on the surface and collects runoff. An unconfined aquifer has a floor but no ceiling, so it receives water as it infiltrates from the surface. The water table of an unconfined aquifer will fluctuate according to the rate that water arrives or leaves.

Like water on the surface, groundwater is always in motion. From the surface it descends slowly through the soil, pulled down by gravity and capillary action. Once it has reached the zone of saturation at an aquifer, it flows in motion known as *deep circulation*, following the aquifer downhill, seeking fissures it can slip through to sink even deeper. How quickly it flows depends on the slope of bedrock and the composition of the soil within the aquifer, but it is almost certain to be very slow. The water in an aquifer is likely to travel only a few feet in a year, a mile in a century.

The rate that water infiltrates the soil and recharges the groundwater supply depends on how dry the soil is and how abundant and fast the precipitation falls. Hard rain quickly saturates the surface layer of soil and causes most of the subsequent rain to run off the surface, while steady rain soaks deeply into the ground. The rate of infiltration depends too on the gradient of the land on which the rain falls (water runs more quickly off steeper gradients), the relative humidity of the atmosphere (dry air evaporates water before much of it can be absorbed into the soil), the nature of rock strata (tipped rock allows water to sink underground while horizontal rock holds it up), and the types of soil and rock in the ground. As water infiltrates downward it follows intricate labyrinths around and between particles. The size of the spaces, or pores, varies with the composition of the soil. Coarse soils have wide, easily traveled pores through which water passes quickly, while dense soils have such narrow pores they resist infiltration. Geologists can judge the porosity of a soil by measuring the ratio of empty space to total volume. Dense igneous rocks, for example, might contain less than 1 percent space and are nearly watertight, while a soil composed of coarse gravel and sand contains about 40 percent space. The pores between particles of clay are so tiny that water tends to get stuck in them by molecular attraction.

Plants are an important factor in the amount of water that infiltrates the ground. A forest or prairie can absorb a great deal of rain because foliage scatters raindrops and slows its descent (allowing as much as 20 percent of the rain to return immediately to the atmosphere as evaporation) so that it does not compact the soil beneath. As much as 95 percent of it infiltrates, flowing down the sides of plant stems and trunks, which serve as entry chambers into the soil, or soaking into layers of decaying vegetation on the surface. Roots hold the soil together to prevent erosion but also loosen it to allow greater infiltration.

When a soil has been stripped of its plant cover and left exposed by clearing and tilling, plummeting raindrops fall like tiny bombs, exploding clods of soil and compacting the surface. Pores become clogged, preventing infiltration and turning the surface to mud. That is why unplanted fields are often covered with standing puddles after rainstorms, and why so much of a farm's topsoil can end up in streams and rivers.

Rain that cannot infiltrate the soil either puddles until it evaporates or runs off. There are no other alternatives. A gentle slope produces a gentle runoff, especially if the slope is covered with vegetation, the stalks acting like tiny baffles to slow and divert the water into millions of miniature rivulets that make steady but relatively slow progress downhill. If the surface is smooth and uniform and the rainfall not excessive, the runoff can take the form of an evenly distributed *sheet flow*. But if the rain is prolonged or heavy—a gully-washer— the runoff scours and gouges soft materials from the surface, carving rills into channels into gullies. As gullies proceed downstream they combine with other gullies to form the dendritic pattern so characteristic of river systems. Most gullies eventually find their way to rivers.

The Phoenicians, Greeks, and Romans all engaged in cycles of overforesting, overtilling, and overgrazing that denuded and eroded the Mediterranean region. The result was described vividly by Homer in the Iliad: "Many a hillside do the torrents furrow deeply, and down to the dark sea they rush headlong from the mountains with a mighty roar, and the tilled fields of men are wasted."

About half of a rainfall will run off a compacted field, eroding topsoil and leaving less water available in the soil for crops. Excessive runoff causes the level of the water table to decline from high ground, creating arid conditions for agriculture, and to rise in bottomlands, causing more frequent flooding and making farming difficult or impossible in the low, fertile regions where it might otherwise be most productive. Excessive runoff causes flooding and silting of streams and increases the volume of pesticides and fertilizers that find their way into our streams, lakes, and oceans.

Though hidden beneath the surface, groundwater remains vulnerable to abuse. It can be contaminated by pesticides or industrial chemicals that infiltrate without being noticed until many years later when the pollutants show up in drinking water pumped to the surface. Organic pollutants can seep into the groundwater from landfills, sewers, and underground fuel tanks, or can run off the surface of highways and parking lots and cultivated fields and infiltrate slowly underground. The same pollutants in a lake or river, though they might cause temporary damage, are eventually cleaned up by microbes. But underground there are few microbes because most groundwater does not contain enough dissolved oxygen to support them. When groundwater becomes contaminated it tends to stay contaminated.

In many places the worst contributor to groundwater pollution is agriculture. Heavy use of fertilizers deposits large amounts of nutrients in the soil. When the three basic plant nutrients—nitrogen, phosphorus, and potassium—enter surface water they cause blooms of algae which die and decompose, robbing the water of oxygen and eliminating much of the plant and animal life the lake can support. But when nutrients infiltrate the soil and migrate downward until they enter the groundwater supply, they become a risk to human health. Nitrates, for example, a common contaminant of aquifers in agricultural areas, are known to cause a type of "blue baby" syndrome that is sometimes fatal in infants. One study found that 39 percent of the wells in heavily farmed South Dakota contained unsafe levels of nitrates. Another study concluded that nitrates from fertilizer can continue to leach into groundwater 80 years after it was applied to a field.

In some regions frequent irrigation of farmland raises the water table, contaminating soil and groundwater with salt. In Australia, salt raised in the spray of surf along the coasts has been carried inland for millennia, where it mixes with water droplets and falls to the ground as slightly salty rain. The rain and its cargo of salt infiltrate the soil, filtering downward into the root zone of acacia and eucalyptus trees. The plants extract most of the water and leave the salt, which continues to infiltrate the soil, eventually accumulating as a concentrated briny solution deep in the subsoil. When Europeans began settling Australia in the 1850s they cleared many of the forests and converted them to pastures and wheat fields. What nobody could have foreseen was that replacing dense stands of trees with low, shallow-rooted crops would allow a much greater amount of each year's precipitation to infiltrate the soil. Over the next several decades that additional water percolating deep into the soil reached an impervious layer of stone and formed a water table. As the water table climbed toward the surface it liberated the reserves of briny solution that had been accumulating underground for so long. By the middle of the twentieth century the water—and the salt—began to appear on the surface. At first farmers and ranchers thought these "saline seeps" were isolated incidents. But they increased in size and number until large expanses of once-fertile land had been poisoned. Then the salt-encrusted fields drained into rivers, increasing salinity until it killed fish in the water and plants along the banks.

Similar saline seeps were noticed in North America about the same time they showed up in Australia. They are especially common in parts of Montana, the Dakotas, and Canada's prairie provinces of Manitoba, Saskatchewan, and Alberta. The North American seeps are unlike the Australian in that they originated, not from sea spray, but from subterranean salt deposits left behind from ancient seas. But like the Australian seeps, they have continued to grow in number. To date they have halted farming and grazing on some 2.5 million acres of land.

Attempts to eliminate the problem of saline seeps have concentrated on two approaches: planting deep-rooted crops such as alfalfa, which use a great deal of water and prevent infiltration below the root zone, thus lowering the water table; and building drainage systems that channel saline water away from the subsoil. Unfortunately, the drain systems have usually been directed into convenient streams, resulting in saline-contaminated rivers— an all-too-common condition in many agricultural regions of the world.

Another risk to groundwater is overexploitation. One of the world's great groundwater reservoirs, the High Plains Aquifer, which includes the

famed Ogallala Aquifer, lies beneath 174,000 square miles of eight U.S. states, Colorado, Kansas, Nebraska, Oklahoma, New Mexico, South Dakota, Wyoming, and Texas and was long considered an inexhaustible source of water. Yet sustained, heavy irrigation of crops such as alfalfa, corn, cotton, sorghum, soybeans, peanuts, and wheat threaten to suck the aquifer dry. According to one study, each day's irrigation draws an amount of water from the aquifer equivalent to the flow of the Colorado River, far more than can be naturally replenished. As a result, by 2011 the water table throughout most of the region had subsided dramatically—by 50 feet or more in a third of the aquifer and by 242 feet in Texas. A 2013 study by the National Academy of Sciences concluded that 30 percent of the groundwater has already been depleted, and at the current rate of usage another 39 percent will disappear by 2063. But few farmers and ranchers are willing to reduce their withdrawals. Many pump water until a well runs dry, then drill another. And the problem is not limited to the American Great Plains. Other aquifers are being similarly depleted beneath Mexico, the Middle East, southern India, the northern provinces of China, and in many parts of Africa.

Aquifers can sometimes be artificially recharged, either by spreading water over the surface or by actually pumping it back into the ground. In Israel, groundwater supplies are replenished by damming flash floods during winter rainstorms and directing the water so that it spreads across sand dunes, where it quickly permeates to the aquifer.

If an aquifer is depleted faster than it can be recharged, it may collapse and never again hold as much water as it originally could. Between 1962 and 2013, the highly irrigated Central Valley of California had lost about 50 million acre-feet of storage capacity in its natural aquifers, an amount greater than the storage capacity of all the artificial reservoirs in California. Collapsing aquifers can create sinkholes or cause large areas of surface above them to subside. Portions of Houston, Mexico City, Beijing, and Bangkok have settled many feet in recent decades. When coastal land subsides in this manner it allows salt water to invade inland, contaminating groundwater and destroying farmlands.

The quality of the water in an aquifer depends on its depth, age, and the soil and rock it is contained in. Fortunately, most aquifers shallow enough to yield water for human consumption are found in sand, gravel, granite, and other fairly insoluble materials. When the water has had contact with easily dissolved stones and minerals, however, its flavor and color are likely to be unpleasantly tainted.

The solubility of water is also responsible for creating caves, especially in regions where limestone is abundant. Limestone dissolves easily in water containing carbon dioxide (dissolved from the atmosphere by rain) and humic acid (picked up when groundwater passes through soil containing decaying vegetation). Limestone is also very hard, which is why it has been widely used for thousands of years as a building material, and it tends to occur in thick, dense, nearly impermeable beds. Infiltrating water seeks any fissures large enough to elbow into, then slowly dissolves the limestone and widens the fissures into cavities. The cavities spread and connect with other cavities and eventually form networks of caves that can extend for many miles. The Mammoth Cave system in Kentucky, for example, has more than 200 miles of charted passageways, and contains vast rooms, including one big enough to contain a nine-story building.

DOWSING FOR UNDERGROUND WATER

The water hidden inside the earth has stirred the imaginations of people for millennia, inspiring much conjecture, fear, and misunderstanding. In a world where water is unevenly distributed, the location of underground supplies is often critical to survival, so anyone with a notion of how to find such water is in demand. In the fifth century B.C. the Greek historian Herodotus described nomadic people in what is now Russia using "divining rods" of willow to find water in the ground. The Chinese may have practiced the same mysterious art as early as 2000 B.C.

Today, the American Society of Dowsers boasts over 3,000 active members worldwide who believe that buried water, oil, gas, or precious metals can be located with the aid of divining rods made of various plastics,

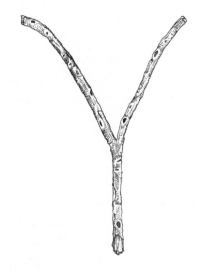

metals, and woods. All that's required for membership in the society is the gift of dowsing, which is said to be possessed by about one in ten people. The usual technique is to hold the dowsing rod loosely in your hands, most commonly with the palms turned up, and walk until an irresistible force causes the rod to twitch or tremble or jump. In some cases the rod will be pulled to the ground with the dowser pulled down after it.

It's been theorized that water and other substances underground could affect magnetic fields and that some people might be sensitive to those slight changes. Most scientists of course believe that water witching is a superstition at best. I'm usually inclined to agree with science, but years ago while working on a construction site I watched a trio of earth-moving machines come to a halt when an engineer suddenly announced that a water main might be buried where excavation was about to begin. Hitting the main would almost certainly rupture it, causing an expensive delay. A young man stepped down calmly from his machine, cut two short lengths of wire from a roll in the back of a truck, bent them 90 degrees, and held them in his hands as if he were pointing a pair of pistols. He then began walking back and forth over the site. After a few minutes the wires in his hand swung inward and crossed each other. He stopped, backed up, and the wires returned to their original position. He stepped forward again and they crossed again. He nodded and another driver shoved a stake into the ground. The two of them repeated the procedure several times, until stakes were arranged in a straight line. Other workers digging with hand shovels found the water main buried four feet beneath the surface—precisely beneath the line of stakes—and construction continued without further interruption.

AQUATIC LIFE UNDERGROUND

We usually think of groundwater as clean, cold, and absolutely lifeless, and we are usually right. Groundwater in most places contains almost no oxygen because it was depleted by organic matter and bacteria as it infiltrated the soil. But water in limestone caves can be oxygenated enough to support a number of animals, all of them well adapted to the perpetual dark of the planet's interior. In caves in limestone country in many parts of the world, including Mexico, Australia, China, Madagascar, and across a swath of the United States from the Ozark and Appalachian regions north to Indiana live fishes that have evolved to live underground. Most are small—up to about four inches long—and are white or pinkish-white in

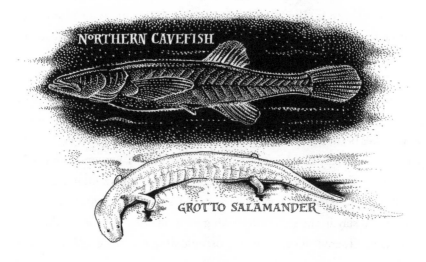

NORTHERN CAVEFISH

GROTTO SALAMANDER

color. All are blind. Some species have no eyes at all but are equipped with hypersensitive lateral line receptors that allow them to detect the subtle vibrations of swimming prey. With their advanced sense of "touch" they are quite successful in hunting such prey as smaller fish and shrimp.

Other cave dwellers include the grotto salamander, which hatches in streams on the surface and while young is equipped with normal eyes it uses to hunt aquatic insects. When it reaches maturity, however, the salamander moves underground, where its eyes grow shut and it hunts by feel alone. The Madison Cave isopod is an eyeless, colorless crustacean found only in caves in portions of Virginia and West Virginia that were first surveyed by Thomas Jefferson.

THE DEEPEST GROUNDWATER

The ultimate groundwater supply might be found as far as 250 miles beneath the earth's surface, in reservoirs formed during the slow, inexorable, and ongoing shifting of the earth's crust. The theory of plate tectonics explains how the pressure of opposing plates forces them to rise into mountains. It works also in reverse. When two plates collide at oceanic fault lines known as subduction zones, one of the plates is forced beneath the other and driven deep into the mantle of the globe. As it goes, it takes ocean water with it. The water is contained in sedimentary rock formed as organic material settled to the bottom of the sea. As the sediment hardens into rock, water is trapped in the pores of the rock itself.

This *connate* water remains trapped in isolated pockets for millions of years. Until recently, most geologists assumed that by the time a plate containing connate water was a few miles beneath the surface all the water would have been squeezed out of it by the weight of the rock above. It would surface occasionally as water vapor in volcanic eruptions, but most of it would be left behind as the mass of rock continued its slow downward plunge.

But some scientists now believe that a significant amount of water manages to stay in the rock, hitchhiking on a ride that takes it hundreds of miles deep. There it remains as microscopic droplets and single molecules

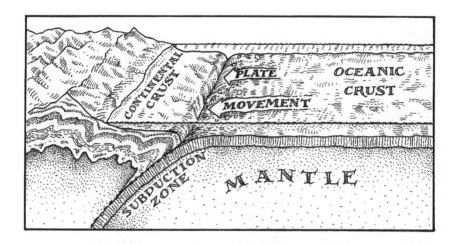

locked away in minuscule spaces in the rock. The evidence for its existence comes from seismic scans used to measure the speed at which earthquake tremors travel through rock of differing compositions. In a region spanning hundreds of miles some 250 miles beneath Poland, the seismic scans have detected rock that seems to contain water. To call it a submerged ocean is a bit fanciful, but a few scientists have suggested that the total amounts involved could equal the volume of all the oceans on the surface.

ARTESIAN WELL

ARTESIAN SPRING

IMPERMEABLE ROCK

GROUNDWATER

IMPERMEABLE ROCK

FAULT

IMPERMEABLE ROCK

4
THE MAGIC OF SPRINGS

If water has magical powers, then the most magical of all must be water newly emerged from the earth. In English the word *spring* is given both to the place where groundwater appears on the surface and to the annual season of birth and rejuvenation, no doubt because water seems to flow from the earth in the same miraculous way that plants spring from the soil. The spring season has the power to restore, renew, and give life, and there is a kind of logic in thinking that springs of new water can cure disease, cleanse the soul, bestow wisdom, and renew the human body.

When Ponce de León set out in 1513 to search for the fountain of youth he was perpetuating a tradition thousands of years old. The emperors of ancient China made frequent pilgrimages to immerse themselves in the waters of springs reputed to restore youth. Alexander the Great launched expeditions in search of such waters. The Wintun and Maidu Indians of California believed they could recover their lost youth by immersing themselves in sacred springs. In Japan it was thought that a rejuvenating fountain was hidden somewhere near the summit of Mount Fuji.

Springs that could restore youth or bestow immortality were described in the folklore and mythologies of many cultures, from Slavic tales of waters that gave life back to the dead, to the fountain of resurrection of the early Christians. Many Asian and European myths described the first creatures on earth being born from springs that watered the trees of life and knowledge. In such traditions springs were fountainheads, upwellings, the springing to existence of something that had not previously existed. They were remnants of the primal waters that once inundated the earth, and thus represented origins in the largest sense, with links to the beginning of the universe and the first seeds of life. The fountains in Rome and the European capitals of the Renaissance were a celebration of water, an ostentatious display of the owner's ability to waste it, and a reconstruction of those original fountains of creation.

Springs could be home to an entire bestiary of mythological creatures, both benevolent and malevolent. Some, like the faeries and pixies of the British Isles, saw to it that no mortals slipped through the waters past the gates of the Otherworld. Others, such as the alluring water nymphs said to live in the spring called Cassotis on Greece's Mount Parnassus, would drag young men into the water and drown them, condemning them to spend eternity as companions to the nymphs. The water nymph Egeria in Roman mythology wept so copiously when her lover died that she became a spring and was thereafter worshiped by pregnant women who thought she could grant easy delivery of their children.

The naiads of Greek mythology were long-lived but mortal water nymphs that inhabited every sizable spring. They had the power to heal and were visited frequently by sick people who drank or swam in their waters. But there was risk involved. Any person who gazed directly at a naiad would be driven insane, and offending one always resulted in misfortune, as the Roman emperor Nero learned when he contracted a fever and temporary paralysis after bathing in a spring. Even now, naiads and nymphs can be found in many freshwater habitats: Aquatic biologists borrowed both names to refer to the immature stages of some aquatic insects.

The earliest wells were enlarged from natural springs and were often the center of community life. A shared well is the hub, not just of a community, but of life itself, for if it goes dry, becomes poisoned, or is in any other way lost, the community is lost. It's not hard to imagine how people could make a leap from community well to holy well or wishing well.

The list of sacred springs and wells is long. The holiest of fountains in ancient Greece was the Castalian Spring at Delphi, where pilgrims washed their hair or bathed in the holy water before consulting the oracle with questions about their destinies. Muslims believe the well in Mecca called Zemzem, or Ishmael's Well, contains the holiest water on earth. The British Isles are dotted with hundreds of ancient wells and springs that were once the focus of pagan practices, were later dedicated to Christian saints, and today are popular with tourists who visit them and toss coins in them as token sacrifices. The Maya of ancient Mexico carried the notion of sacrifice to the extreme by throwing people into Yucatan *cenotes*, or sinkholes, in an effort to appease the gods of the underground water.

Sacred or not, springs are the places where aquifers intersect the surface. If they are small they are known as *seepage springs*. Seeps might trickle with only a few tablespoons of water an hour, but where they are found in clusters they can produce a substantial flow, enough in many cases to form the headwaters of rivers. Larger springs are sometimes known as "boiling springs" because they issue from the ground with enough power to rollick like water boiling in a saucepan.

The most prevalent springs are probably the unseen ones beneath rivers, lakes, and oceans. A spring-fed lake is a kind of surface extension of the groundwater supply, fed by and feeding aquifers extending underground around the shore. A river can likewise be perched on top of an aquifer that gives it a constant supply of new water. When the Manavgat River in Turkey enters gorges in the Taurus Mountains it is a modest stream. No tributaries join it, yet it emerges a few miles downstream as a major river, roaring with volume, swollen by thousands of gallons of water gushing from streams beneath its surface.

Springs that jet under pressure are *artesian springs* if they occur naturally, or *artesian wells* if they are produced artificially. The name is derived from the old province of Artois in northern France, where the first artesian well was drilled in 1126. Artesian springs and wells originate from groundwater sandwiched between watertight layers of rock. If such a confined aquifer is not horizontal it develops a "head" of pressure, with water at the upper end bearing down on the water at the lower end. An artesian spring is where groundwater under such pressure naturally escapes the aquifer; an artesian well is drilled into the lower end of the aquifer, releasing water under pressure.

Artesian springs and wells are found in a number of famous oases in desert regions. A large artesian spring in Lebanon, Ain ez Zarqa, gushes from a fault in deep limestone formations and becomes the Orontes River, Lebanon's most important river. The ancient oasis known since Roman times as Nepte is considered one of the most beautiful and bountiful in the northern Sahara Desert. At its heart, amid orchards of date palms, lemons, and oranges, are dozens of bubbling artesian springs.

Many springs are relatively stable, discharging approximately the same amount of water every season because the slow filtration of groundwater tends to distribute the supply evenly. In limestone regions, however, springs sometimes ebb and flow under the influence of natural siphons in subterranean chambers, and are known as *periodic springs*. The appropriately named Periodic Spring, in the Bridger-Teton National Forest in Wyoming, surges with about 200 gallons of water per second for periods ranging from 4 to 25 minutes, then goes dry for about the same amount of time. The cause of these periodic surges is a subterranean S-shaped passage that leads from a reservoir deep in the limestone bedrock. Water from rain and snow infiltrates the limestone miles away and gradually finds its way to the hidden reservoir. When the reservoir has filled above the upper elbow of the S, the passage becomes a siphon. It floods, pushing air out, and expels water from the spring mouth.

When a limestone region has become permeated with networks of caves and underground streams it becomes a *karst* topography, named for a plateau in the Dinaric Alps northeast of the Adriatic Sea. The only two requirements for such a region is that it have extensive deposits of limestone and a wet climate. The combination of rainfall and soluble bedrock creates a number of landforms typical of karst topography, including sinkholes (caused by the collapse of cave ceilings), solution valleys (formed when sinkholes expand and merge into large depressions), and disappearing streams that flow along the surface then pour down sinkholes into underground channels, often emerging, sometimes many miles away, as springs. In North America, well-known karst regions are found in Florida, Kentucky, southern Indiana, and on Mexico's Yucatan Peninsula, where the abundant cenotes once worshiped by Maya are now an important source of drinking water. Elsewhere, karst topography is found in the Slovensky Raj ("Slovak Paradise") of the eastern Czech Republic, in southern France, and in southern China, where a spectacular landscape is dominated by steep, loaf-shaped towers of rock left behind when softer limestone eroded.

Ponce de León's fountain of youth may not exist in Florida, but many

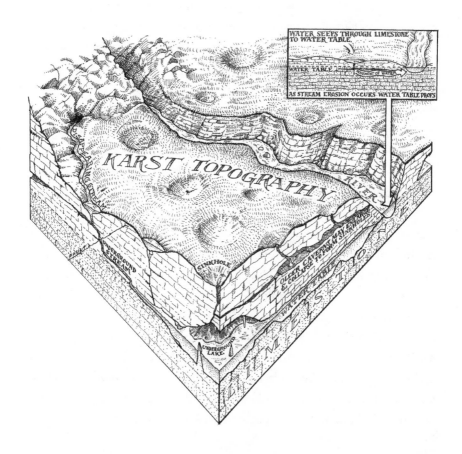

other fountains do: Florida is home to perhaps the greatest number of natural springs on earth. Among the largest of them are Silver Springs, which gushes 800 million gallons each day from 150 outlets; Rainbow Springs, which produces 421 million gallons per day; Itchetucknee, with 300 million gallons; and Wakulla Springs, with about 210 million gallons per day.

Limestone formations in the Ozark Mountains also produce many springs. The best known of them is Big Spring, which flows from a collapsed cave a short distance from Van Buren, Missouri, and pumps up to 840 million gallons of water every day into the Current River. Nearby Blue Spring—named for the color loaned it by mineral content, clarity, and depth—discharges 90 million gallons a day into a gently boiling pool 100 feet across and more than 250 feet deep.

When a karst region is surrounded by impervious rock, groundwater can escape only from springs in low-lying areas, sometimes giving birth to entire rivers from a single source. The famous La Fontaine de Vaucluse, near Avignon, France, is one of the world's great springs. Described and celebrated by writers since Pliny and Petrarch, it flows from a small, downward-sloping cave at the base of a 600-foot-high cliff and is the source of the Sorgue River. During normal flow it pumps about 400 cubic feet per second—about 260 million gallons each day. But at its peak, during the spring, it can exceed 10 times its normal flow. The maximum flow measured at Vaucluse is more than 3.8 billion gallons per day.

Another famous spring, Goueil de Jouéou, is the sole source for the Garonne River in southwest France. The origin of the spring's water was debated for centuries. Some thought it came from a large sinkhole known as Trou du Toro, "the bull's hole," located on the south slope of the Pyrenees, a mountain range away. The sinkhole earned its name from the roar it made when meltwater poured into it and disappeared underground. Most people assumed the water flowed from there south toward the Ebro River and the Mediterranean. But when the Spanish proposed to divert water from Trou du Toro to operate a hydroelectric plant, the French financed research to learn the ultimate destination of the water. Researchers dumped green dye into Trou du Toro and were gratified to see it emerge several miles north, on the other side of the mountain, as a green cascade from the spring of the Garonne River. The Spanish were persuaded to let the water flow.

Also in southern France is the spring known as Les Bouillens, "the bubbling waters," where Perrier springwater is bottled. Every day the spring emits 500,000 gallons of naturally carbonated water. The effervescence results from carbon dioxide that rises through porous limestone from volcanic sources several miles below the surface. As the carbon dioxide rises it dissolves in the aquifer.

When springs emerge from karst regions near ocean coasts they often flow a short distance as open rivers and empty into the sea. They also enter the sea directly, as submarine springs, a few of which are large enough to be conspicuous on the surface even miles from shore. In the Gulf of La Spezia in the northern Mediterranean a spring rises through 60 feet of saltwater and heaves the surface with a small hill of freshwater. Another large freshwater spring bulges the surface of the Gulf of Argolis, in southern Greece. A similar spring a couple miles off the coast of Cuba in the Gulf of Batabanó creates so much turbulence that small boats can approach it only with caution. For centuries sailors and fishermen have used such offshore springs to replenish their supplies of drinking water.

Springs are usually regarded as the purest and safest source of water. After being filtered through miles of sand, gravel, and rock in oxygen-depleted aquifers, water tends to become very clean. That's not true everywhere, of course—it is probably a bad idea, for instance, to drink from the seeping springs that can be found throughout Manhattan's Central Park—but in many parts of the world, if the water in a spring is cold and clear it is probably safe to drink. It may not shower you with good fortune or heal what ails you, but it is guaranteed to be a satisfying beverage on a hot day.

LIFE IN SPRINGS

The life that grows up around the constant dampness of a spring can be the most diverse ecosystem for miles around. Such communities of plants, insects, and vertebrates often remain stable, even while surrounding regions go through great changes, and may include relic species left behind from past ages.

Yet the spring that is crowded to its edges with so much life is often barren of life itself. When water first escapes from the ground it is usually so depleted of oxygen that only a few bacteria can live in it. Only after it has tumbled over ledges and riffles does enough oxygen get stirred in to make it habitable to most aquatic organisms.

Not all springs are lifeless, however. Many Florida springs are occupied by species of mayfly larvae that need little oxygen to live and a few top-dwelling minnows that survive by gulping air at the surface. Throughout Europe are several species of flatworms, snails, stoneflies, a flightless beetle, and a few other organisms that have colonized the oxygen-depleted water of springs. Springs in Denmark are often inhabited by a species of caddisfly that probably originated in the Arctic and was left behind after the last ice age. Such relic species are more often found in old springs than new ones.

There are fewer in Lapland than in the Alps, for instance, and fewer still in recently glaciated North America. The exceptions in North America are a scud (a small shrimplike crustacean) found in a single New Mexico spring and in others in northeastern North America. A scud and several sowbugs and water lice are found in only one spring in Doe Run, Kentucky.

Some springs and creeks in the American Southwest are inhabited by pupfishes, small members of the killifish family named for the frolicsome courtship behavior of the males. The pupfishes live in some of the world's smallest and most isolated habitats that can support vertebrates: swimming-pool-sized springs fed by deep aquifers. Biologists have likened those desert springs to the Galápagos Islands and other pockets of habitat where isolated populations evolve into species. After 40,000 years of speciation and adaptation, the little pupfishes have become extremely hardy. They can tolerate water with temperatures over 100 degrees Fahrenheit (and can survive for short periods at temperatures up to 113) and in salinity several times that of the oceans. Unfortunately, their confined habitats make them vulnerable. The White Sands pupfish is an endangered species found in only two springs and a small stream on the U.S. Air Force's White Sands

Missile Range, in southern New Mexico. The Devil's Hole pupfish, a one-inch-long, iridescent-blue fish that is one of the world's rarest fish, lives only in a 10-by-50-foot pool of 93-degree temperature in a submerged cavern in Nevada's Mojave Desert. As of September, 2013, only 65 of the tiny fish were known to be alive. It was the smallest number counted since 1976, when a Supreme Court ruling established Devil's Hole as a National Monument to protect the water and the pupfish.

DEVIL'S HOLE PUPFISH
Cyprinodon diabolis

SURFACE VENT

GROUNDWATER PERCOLATES DOWN INTO HOT ROCK

HEATED WATER IN CAVERNS AND FISSURES RISES AS STEAM AND FORCES WATER UPWARD

HOT ROCK

MUDPOT

5

IN HOT WATER: THERMAL SPRINGS AND GEYSERS

The earth is well insulated. Store food in a cellar only a few feet deep and it stays about the same cool temperature summer and winter. Even during the coldest winters the ground rarely freezes more than two or three feet below the surface. Dig down about 25 feet and the temperature remains constant at the average annual temperature of the air in that region.

But dig deeper and temperature rises. Typically it rises about one degree Celsius for every 80 feet closer to the molten core of the planet. Temperature rises much more rapidly in those places on earth where that molten core has slipped close to the surface. It is a strange sensation to stand on hot ground in Iceland, New Zealand, or Yellowstone National Park and feel the planet's overheated machinery shuddering beneath your feet.

Water underground can be warmed slowly by the residual heat of the earth or heated quickly by contact with stones in proximity to magma and the chambers of old volcanoes. Such chambers can remain hot for thousands or even hundreds of thousands of years and they make very efficient cookers of water. When heated water finds a way to the surface it spills out as a *thermal spring*, or, if under pressure, as a *geyser*.

The standard definition of a thermal spring is one with water temperature more than 10 degrees Celsius above the yearly average of the surrounding air. Otherwise it differs little from an ordinary spring. If the water in the spring contains more than approximately one part per thousand of dissolved substances (about the amount that gives water a noticeable flavor, scent, or discolored appearance) then it qualifies as a mineral spring. Most thermal springs are rich in dissolved minerals, and most mineral springs are warm or hot.

All geysers have certain features in common. They are found in regions where hot rocks lie fairly close to the surface; where irregular fractures in the rock extend from the surface to a subterranean region of intense heat; and where a large supply of groundwater is available. As water trapped in fissures and caverns is heated, it reaches a balance of temperature and pressure that leaves it poised to become steam. Because water deep in a geyser system is under pressure, much like water in a stove-top pressure cooker, its boiling point rises. When it finally boils, it creates expanding steam, which shoves the water above it out of the geyser, shooting a spray of water and steam into the air. Once this upper water is ejected, the pressure below drops, allowing the remaining water to convert to superheated steam, causing a second violent eruption. When the eruption ends, the geyser settles down for a time while a new supply of water percolates into the fissures below and begins to heat. This intermittence is characteristic of geysers all over the world. Intervals between eruptions can be regular or irregular, and can vary from a few minutes to many weeks.

Our word geyser is adapted from Geysir, "the Gusher," a once-spectacular geyser in Iceland that became famous throughout Europe during the Middle Ages. Geysir's power has diminished with age and it is now overshadowed by nearby Strokkur, "the Churn," which erupts every four to ten minutes to heights up to 100 feet. It and a number of smaller "sister" geysers are features of an extensive thermal area in southwestern

Iceland, where molten magma escaping through the Mid-Atlantic Ridge comes near the surface of the island.

Where there is volcanic activity, there are usually hot springs and geysers. The Japanese islands contain about 20,000 hot springs. The Valley of the Geysers, on Siberia's Kamchatka Peninsula, is scattered with more than twenty geysers, the largest of which erupts every hour and spews water and steam up to 500 feet high. Soborom Hot Springs, in northwestern Chad, are a cluster of hot springs and geysers on a volcanic plateau above the Sahara Desert. The Larderello Hot Springs, located southwest of Florence, Italy, include vents that spew steam as high as 165 feet in the air and were thought to have inspired Dante's vision of Hell in the *Inferno*.

Nowhere is there a larger number and greater variety of geothermal features than in Yellowstone National Park in northwest Wyoming. This oldest of U.S. national parks (established in 1872) contains about 300 geysers and 10,000 hot springs, steam vents, mudpots, and other thermal outlets within its 3,500 square miles. Close beneath the surface of the park is a reservoir of molten rock left behind after the last of the region's volcanic eruptions 100,000 years ago. When rain and melting snow infiltrate the ground in Yellowstone, the water enters aquifers that make frequent contact with hot stones. The heated water returns directly to the surface as hot springs, or, if under pressure, as some of the most famous geysers in the world.

The best known of all is Old Faithful, which, true to its name, erupts about every 70 minutes, night and day, year after year. During each eruption, steam and hot water jet for five minutes to an average height of 130 feet and a maximum of 184 feet. Other Yellowstone geysers vary in the power and duration of their eruptions. The highest of them was Steamboat Geyser, which until it ceased erupting in 1969 reached heights of 400 feet. Great Fountain and Grand Geysers spout as high as 200 feet. Riverside Geyser sends a lower but very broad plume of spray and steam arching over the Firehole River every 5 hours and 45 minutes.

Yellowstone's hot springs are found in every size and configuration, from tiny mudpots to pools the size of small lakes. Some are lukewarm, some boiling. The most dramatic are Mammoth Hot Springs, where hot water spills over terraces and stairs that make the formation look something like a gigantic wedding cake. An estimated two tons of dissolved limestone comes to the surface of the spring every day; as the water cools the limestone precipitates on the edges of the spring, adding to elaborate terraces. Other hot springs, like Grand Prismatic Spring and Morning Glory Hot Pool, are notable for the colors produced by algae and bacteria. White, yellow, mother-of-pearl, indigo, sapphire, brown, green, and red give the water and the edges of the pools the multicolored hues of ice cream at an intergalactic carnival. The mudpots of Yellowstone are hot springs that occur in clay. The blue, pink, white, yellow, and pastel blends in Fountain Paintpot and others are caused by iron compounds in the clay.

Geysers by their nature tend to be short-lived. As dissolved minerals precipitate around the rim and hot water erodes the chambers beneath it, a geyser can become an ordinary hot spring or can become clogged altogether, although new ones are likely to appear in the same area. The Waimanugu Geyser on New Zealand's North Island has not erupted since 1908, but in its heyday it was the most powerful geyser on earth, blasting every 34 hours to heights of 1,500 feet.

Hot springs and geysers have been exploited for human use since at least the Roman Empire, when the water from thermal springs was diverted into the city to heat houses and fill public baths. Steam vents were first put to use to turn turbines and generate electricity in 1904, in Larderello, Italy. Since then geothermal energy sources have become widely used in Japan, Russia, Mexico, New Zealand, the United States, and Iceland (where the capital city of Reykjavík is heated almost entirely with hot water piped from underground reserves). The largest geothermal site in the world is at the steam field known as the Geysers, located about 70 miles north of San Francisco. Steam-driven turbines there have the capacity to supply San Francisco with about half of its electricity needs.

Long before hot springs and geysers were put to work powering machinery and generating electricity, they were popular for soothing tired and aching muscles. People have been slipping into hot baths since the first daring hominid stuck a tentative toe into a steaming pool. They were attracted no doubt to pools like the one my wife and I once visited every evening for six weeks beside the Madison River in Yellowstone Park. It was a small, steaming, slightly sulfurous pool of bathtub-perfect water, 100 degrees Fahrenheit, just a pebble's toss from the cold and strong-running Madison. At the end of a day of hiking and fishing we would ease into the gently bubbling water, lay back with our heads against the bank, and watch the stars come out. No Roman emperor ever had a finer bath.

Taking the Waters

Hot, mineral-rich springs have long been valued for their curative powers. Medicinal springs in North America have given rise to such famous resort communities as Radium Hot Springs in British Columbia; Banff in Alberta; Palm Springs, California; Hot Springs, Arkansas; and Saratoga Springs, New York. Elsewhere in the world, health-conscious bathers congregate at spas in Bagneres-de-Luchon and Bareges in France, Cheltenham in England, Bad Ems and Baden-Baden in Germany, Sedlcany in the Czech Republic, and the ancient Turkish baths in Budapest.

The ancient Romans raised therapeutic bathing to an art form and made it central to the social life of the city. When the first aqueduct was built in 312 B.C., it created such a sensation that there were soon 381 miles of them carrying water to Rome. At one point the average Roman citizen consumed 300 gallons of water per day (in the United States today we each use about 100 gallons for each day's personal and household purposes). The aqueducts could channel nearly 200 million gallons of water into the city's 1,352 fountains and more than 900 hot, warm, and cold public baths. No day in a Roman's life was complete without a few hours spent soaking in the baths.

Public baths enjoyed popularity for centuries after the collapse of the Roman Empire but they fell out of favor in Europe late in the sixteenth century when it was rumored that they were the transmitter of the syphilis epidemic sweeping the continent. Public bathing did not become popular again until the eighteenth century, when physicians began promoting the healthful properties of soaking in hot mineral water. By then, "taking the waters" could mean bathing in hot springs or drinking mineral water or seawater. Even the air in the vicinity of such waters was thought to be healthful.

A vivid account of taking the waters was left by Madame de Sévigné, whose letters to her daughter provide an insightful look at upper-class life during the reign of Louis XIV. A letter dated June 4, 1676, describes taking the medicinal waters at Vichy:

> At length I have finished douching and exuding; in one week I have lost over thirty pints of water, and believe myself to be immune from the rheumatics for the rest of my life. There is little doubt the cure is a painful one, but a wonderful moment dawns at last when, empty and renewed, one sits relishing a cup of fresh chicken broth which joy is not to be despised, indeed I rank it very high: this is an adorable place... Tomorrow I take a mild dose, drink the waters for another week, and the trick is done. My knees are as good as cured, I still cannot close my hands; but using me as a bundle of soiled linen has proved a highly successful operation.

*

Bathing your ills away is not necessarily an unsound medical practice. Water rising through layers of rock dissolves the minerals it comes in contact with and brings them to the surface. When Roman physicians prescribed water cures for the treatment of emotional disorders, they had no way of knowing that a common ingredient of mineral springs is the element lithium, which is now widely used in the treatment of manic-depression. Spas and springs that for centuries were said to clear up skin conditions often contained sulfur, a common ingredient in many of today's acne medicines. Waters with a high content of dissolved sodiums or sulfates of magnesium act as diuretics and purgatives, and were certain to cleanse the body of poisons and excesses. And if the waters cured heartburn and indigestion, as was often claimed, perhaps it was because they contained sodium bicarbonate, the miracle ingredient in baking soda and Alka-Seltzer.

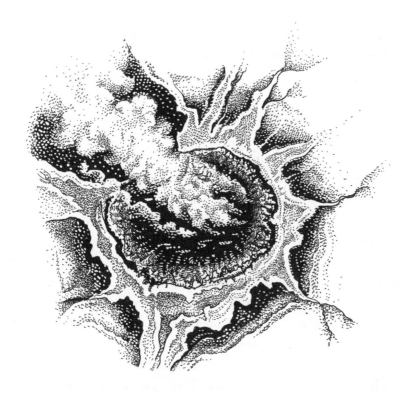

When the Romans conquered Europe they claimed most of the major hot springs for their own. In A.D. 76 they took the "King's Bath" at the center of Bath, England, which had already been a religious shrine and medicinal spa for at least 5,000 years, and replaced the ancient Celtic shrine to the god Sulis with a shrine to their goddess Minerva. The spring is one of the most copious in the British Isles, releasing about 250,000 gallons of 120-degree water each day. In the sixteenth century it was visited by John Leland, chaplain and librarian to Henry VIII, who observed (in the haphazard spelling of the day) that the bath was "much frequentid of people deseasid with lepre, pokkes, scabbes, and great aches, and is temperate and pleasant."

Europeans eventually became connoisseurs of hot mineral springs, identifying three basic types. Saline springs included those containing mostly magnesium sulfate and Epsom salts—named for the mineral springs in Epsom, England—and were valued for their purgative powers. Springs containing iron oxide were thought to have restorative powers. And springs emitting the unmistakable rotten-egg odor of sulfur or hydrogen sulfide were valued for their benefits to the skin.

Baden-Baden ("Bath Bath"), Germany, was popular among Romans as early as A.D. 117 and continues to be among the best-known of the European spa resorts. Its various mineral waters contain iron, manganese, selenium, lithium, and arsenic, and have been credited with curing most of the usual physical ailments plus general sluggishness of spirit and melancholia. The largest spring there, the site of the Caracalla Spa, releases 500,000 gallons of medicinal water a day.

The town of Spa, Belgium was the first to use the word *spa* to describe its hot springs. The word is thought to be either an acronym of Nero's legendary statement "Sanitas per aquas" ("health through water"), or was derived either from the Walloon word *espa* ("fountain") or the Latin *spargere* ("to scatter, sprinkle, moisten"). The town of Spa opened as a resort in 1326 to capitalize on a steady stream of pilgrims who had been arriving for decades convinced the springs could perform miraculous healings. It soon became the center of a cult following of St. Remaclus, who was thought to have the power to purify fountains and make new springs appear. By 1351 so many pilgrims were visiting Spa that the town levied a "cure tax." In later centuries, it became a popular destination for royalty from many nations, as well as such literary and intellectual luminaries as Victor Hugo, Freud, Gogol, Nietzsche, Mark Twain, and Casanova.

Native Americans made hot springs the centers of various religious rites, often attributed healing powers to them, and sometimes declared them neutral ground where warring tribes could rest and recuperate in peace. Among the thermal springs considered sacred before the arrival of Europeans were the springs in the Bow River Valley of Alberta, Boyes Hot Springs in California, Pah Tempe Hot Springs in southwestern Utah, French Lick Springs in Indiana, Saratoga Springs in New York, and White Sulphur Springs in West Virginia.

The Natchez, Cherokee, and other tribes believed a benevolent god lived in the waters now known as Hot Springs, Arkansas, and that at the birth of the earth the first man and woman were born from those steaming waters.

By the late nineteenth century Hot Springs was immensely popular among people seeking relief for bodily ills. A report in 1885 listed a nightmarish cornucopia of ailments that could be cured by the 140-degree mineral waters. They included gonorrhea, syphilis, rheumatism, lupus, measles, ulcers, blood poisoning, eczema, gout, boils, enlarged glands, tumors, and "diseases of the womb." In the early days of the resort community, the

individual springs were assigned special healing powers. Depending on what ailed you, you could immerse yourself in waters specialized for the feet, the stomach, the kidneys, the livers, and the skin. One favorite spring was for staving off old age.

Many springs have been said to grant procreative powers, perhaps none more so in modern times than a cold mineral spring in the Balkans, near the town of Kladanj, between Belgrade and Sarajevo. The spring became famous in 1969 as the source of mineral water known as *muska voda*—"man's water"—which local shepherds had been giving to their rams for years because they believed it caused them to father more lambs. Word of the water's special powers spread after a journalist interviewed an 87-year-old man who credited the water with allowing him to father 21 children, the last one born that very year. Soon it was bottled and sold commercially and for a time was in high demand.

No nation has more thermal springs than Japan, nor, perhaps, as many inhabitants who believe in their therapeutic advantages. The 3,500 springs and geysers near Beppu make that city one of Japan's most popular vacation destinations. The springs with the most medicinal value are almost unbearably hot and are known as *jigokus*, or "hells." They are ranked according to their colors, which vary with mineral content: blue, vermilion, emerald, yellow, and clear-white. Perhaps the most famous of Japan's hot springs are those in the mountain forest of Yamanouchi, in the central Japanese Alps, where Japanese macaques can be seen almost any day—especially in winter—soaking in the hot water, their bright red faces like the faces of well-steamed humans.

Taking the waters continues to be practiced on a large scale. Anyone who vacations at a spa, drinks a bottle of mineral water, or soothes a pulled hamstring in a whirlpool bath is upholding a tradition of water therapy that was advocated by Hippocrates as early as the 5th century B.C. In 1968, when Roy Jacuzzi invented the first portable hot spring and demonstrated it to a skeptical audience at a county fair, he was promoting hydrotherapy, not a stylish accessory for suburban decks. He claimed that a daily soak in the churning water of his tub was relaxing and healthful and could cure many of the minor aches and pains of life. It took a few years to catch on, but people eventually got the idea.

LIFE IN HOT WATER

We boil water to kill organisms, so it's natural to assume that hot springs are lifeless. But even there life flourishes. Some of the hottest of Yellowstone's springs have been colonized by algae and bacteria that give the springs their characteristic colors. The water also contains many microbes that have adapted to the heat. Those single-cell survivors are so hardy that researchers are examining them to see if they can be used to get rid of nuclear waste or chew the paint off old ships. One such microbe has already proven useful by providing a chemical that makes it easier and many times faster for scientists to analyze DNA fingerprints.

Hot springs are home also to some of the hardiest of insects. The Ephydridae is a family of about twenty species of highly adapted flies sometimes known as shore flies. Some, the brine flies, live in highly saline waters like Great Salt Lake. One species lays its eggs in pools of oil and petroleum and feeds on insects that have fallen into the deadly stuff and died. Others deposit their eggs in springs as hot as 112 degrees Fahrenheit. Since hot water contains almost no oxygen, the larvae must live near the surface, where they feed on floating algae and breathe air through retractable breathing tubes much like tiny snorkels. The adult flies hover around the edges of the springs and feed on less heat-tolerant insects that have been unfortunate enough to fall into the water and get parboiled.

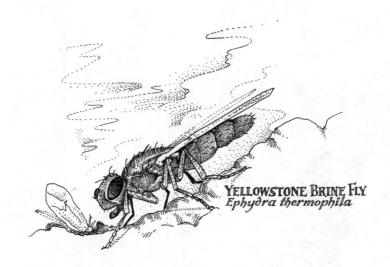

YELLOWSTONE BRINE FLY
Ephydra thermophila

6

THE BIRTH OF A RIVER: BROOKS AND CREEKS

Water is ageless, both ancient and young at once, but the places where it gathers can show their years. The oceans are old, and look it. Big rivers often have the dirty and timeworn look of roads rutted by years of hard use. Bogs with their stained and stagnant water are old lakes, bearded and decrepit, already with something of the grave about them. But creeks and brooks are young. They're the kicking colts of the aquatic world, newborn and clean and fresh, tripping down mountain slopes, frolicking across meadows and woods, galloping ahead to meet whatever the future brings. Cross paths with a creek and it's likely to bring out the kid in you. You can't help getting your feet wet.

Depending on where you are, a small stream can be a rill, gill, rillet, runnel, rivulet, creek, crick, brook, branch, sike, beck, or burn. *Random House Unabridged* says a *brook* is the smallest natural stream of freshwater, *creek* is a stream larger than a brook and smaller than a river, *river* is a fairly large stream flowing in a definite channel, and *stream* is any flowing body of water, whether it be brook, creek, or river. In practice those definitions are rarely absolute. Often a name is a matter of whim or regional convention. Idaho's Silver Creek, for instance, is considerably larger than Michigan's Boardman River.

Early in the twentieth century, biologists in western Europe devised a stream classification system based on the fish that lived in them. Upper reaches were labeled the trout zone, followed downstream by grayling, barbel, and bream zones. This system was of course valid only in that relatively small region of the world where trout, grayling, barbel, and bream were found, and it tended to fall apart in river systems where impoundments, lakes, and other features made it possible for trout to live in lower reaches and bream to live in upper. Efforts to be more specific included classification based simply on width: brooklets were less than a meter wide, brooks were 1 to 5 meters, little rivers were 5 to 25 meters, rivers were 25 to 100 meters, and large rivers were 100 to 300 meters.

In an 1896 short story titled "Crocker's Hole," British writer R.D. Blackmore noted a definitive but largely unacknowledged distinction between rivers and smaller streams. "In the Devonshire valleys," he wrote, "it is sweet to see how soon a spring becomes a rill, and a rill runs on into a rivulet, and a rivulet swells into a brook; and before one has time to say, 'What are you at?'... we have all the airs and graces, demands and assertions of a full-grown river. But what is the test of a river?... 'The power to drown a man,' replies the river darkly."

Most aquatic ecologists today agree with Random House in designating any mass of water moving through a defined course a stream, but are likely to label rivulets as the smallest streams and brooks as the next largest. Rivulets flow into brooks, which flow into creeks, which flow into rivers. When a more specific system is needed, streams are sometimes ordered according to the number of tributaries they gather. A first-order stream flows directly from a source, a second-order stream has one tributary, a third-order stream has two tributaries, and so on. Yet another system is to classify streams by a combination of criteria such as size and bottom type. Thus small streams less than 10 feet wide are *headwater streams*, and can be either rocky creeks or *marshy creeks*. Larger streams, more than 10 feet wide, are either mid-reach streams (those with conspicuous pools and riffles) or *base-level streams* (sluggish, meandering streams with occasional sandbars but few pools and riffles). Streams wide enough and deep enough to be navigated are simply *large rivers*.

Small streams are among the most inviting forms of surface water. They can be easily explored on foot and with equipment no more complicated than a pair of hip waders and a specimen net. Some headwater streams flow for miles, but many are short enough to be explored in a single day. And though it can require much time and effort to discover the source of a river—it took 2,000 years to find the Nile's—the places where creeks are born are often just a short hike uphill.

LIFE IN SMALL STREAMS

In my part of the world the best time to explore creeks is in April and early May, after the spring runoff has subsided and before mosquitoes hatch and trees sprout with leaves. One recent May afternoon I walked along a stream I already knew well in its lower reaches, where it flows through a valley of cedars and hardwoods before merging with a famous trout river. That lower water is strong and cold and dark, too wide to jump across but still small enough to justify calling it a creek.

It is brook trout water, and I was fishing it with a fly rod rigged with a short leader, a splitshot, and a wire hook baited with a worm. Faint paths had been worn along the banks by other anglers, human and animal. I fished slowly upstream to unfamiliar water, passing though cedar groves and small hidden meadows, and around a long stretch of unfishable water choked with tag alders. The creek grew smaller and colder, the pools shallower and less promising. I caught trout—brilliant seven-inchers that came out of the water gyrating so madly I could not get my hand on them—but they were fewer than downstream and smaller. When the creek grew too shallow for even small trout I stopped being a fisherman and became an explorer.

The valleys where creeks flow tend to be damp places with rich soil and much vegetation. Typically, the bottomland is filled with tag alders, willows, osiers, and other moisture-loving shrubs and small trees growing in dense and intertwined thickets. Anglers tolerate those thickets with fondness because they make effective guardians of some of our most productive trout streams.

Not every stream, of course, is well guarded. Mountain streams flow through high meadows bright with wildflowers or glissade down corridors of rock as wide and flat as paved roads. Desert streams flow through open canyons and sandy arroyos, exposed to sun and wind. Elsewhere, it is possible to find woodland creeks weaving among mature cedars or pines that have crowded out the lesser trees to create an open, canopied environment as accessible as a city park, with mossy fallen trees to serve as benches and game trails paralleling the creek like footpaths.

Creeks are likely to be predominately rocky or marshy. Rocky creeks are fed by springs, runoff, melting snowpack, glaciers, or a combination of them. They can originate in springs so consistent that their flow varies only a few inches from highest water to lowest, or they can be seasonal and intermittent, flooding after storms and reduced to trickles between rains. They are often steep and fast, with few pools, and follow a relatively straight course through young, V-shaped valleys. The water in these fast-flowing creeks is clear if fed by springs, milky when fed by glacial melt, or dark with sediment if there is much erosion upstream.

Marshy creeks usually flow through low-lying areas and are spring fed. Their bottoms are of mud, clay, soft sand, or organic debris and often support dense growths of aquatic vegetation. They meander slowly through grassy marshes or through lowlands thick with alders and other marsh-loving trees—vegetation that, as it dies, stains the water the color of tea. They will be cold if fed by frequent springs, but their slow current and dark color warms them in the sun and they may be home only to warm-water species of fish, whereas rocky creeks are more likely to support trout and other cold-water species.

Most creeks contain a wide variety of habitats—riffles and pools, slow water and fast water, silt bottom and gravel bottom—which allows them to support a diversity of life unrivaled by many still-water environments. Never mind the microscopic myriads, just those creatures you can hold in your hand and observe with naked eyes make up a community of such complexity and variety it could (it *does*) inspire volumes of research.

The key to the life in a creek is dead things. Energy flows through every aquatic ecosystem, whether creek, pond, lake, or ocean. It begins with green plants, which serve as *primary producers* able to manufacture nutrients from sunlight, water, and carbon dioxide through photosynthesis. In ponds, lakes, and oceans, the most important primary producers are algae. But most algae are plankton (from the Greek *planktos*, "wandering") that drift with wind and current, and are therefore quickly washed from streams. In running water, algae called diatoms grow on the surfaces of rocks, but the most significant primary producers are leafy plants in the water or on the banks beside it. They provide the foundation for the entire network of plants and animals in the stream by supplying fallen leaves, bits of crumbling wood, and pieces of dead aquatic weeds. The importance of such raw material can be seen in mountain streams that flow directly from snow-filled valleys high above the tree line. Though pristine and gorgeous such streams are often nearly devoid of life. If mountain streams originate in forests or wooded ravines, however, where fallen trees crisscross the water and leaves fall into it every autumn, they are probably inhabited by aquatic communities as diverse as those of marshy and woodland creeks.

Big trees are especially important to the life in streams. Small debris is flushed quickly from a stream, but when a sizable tree falls across the water it stays there for decades. It is colonized soon after it dies by boring and tunneling insects that open passages for further invasion by other insects, algae, and microbes. Moisture penetrates the bark and begins softening the inner wood. Fungi take hold and begin the slow work of decomposing tough fibers. Eventually oxygen penetrates the softened wood, followed by earthworms and larvae usually associated with soil. Indeed, at some point the tree becomes soil.

As the tree is slowly consumed, a steady rain of organic material falls to the stream below. If the flow of a stream is strong and unobstructed the organic material is quickly flushed away downstream. But fallen trees and large rocks create tiny dams and pools that slow the current, allowing bits of wood and fallen leaves to settle to the bottom and begin to decompose, releasing nitrogen that feeds algae on the bottom and becoming a feast for bacteria, aquatic fungi, and larvae of such insects as black flies, caddisflies, mayflies, and stoneflies. Some bottom-dwelling insects are "shredders" that tear bits of softened wood into minute pieces and ingest them; others are "raspers" that scrape nutrients from the surface of the wood. Together they succeed in crumbling the wood and other organic debris on the bottom into a drifting compost called detritus that provides food for other organisms downstream.

Organisms that eat vegetable matter are the primary consumers of an aquatic ecosystem. They in turn are preyed on by secondary consumers such as fierce-looking aquatic beetles and the larvae of dragonflies, dobsonflies, certain stoneflies, and caddisflies. These predators prowl among the stones on the bottom in search of prey, but often become prey themselves to tertiary consumers such as trout and other fish. For an angler, the secret to finding fish in a stream is to find the creatures it depends on for food.

It seems logical to assume that small streams are inhabited only by small fish, but that is not always true. With sufficient food and cover, small waters can support surprisingly large fish. Even relatively barren headwater streams are often used for spawning by mature trout, salmon, and other fish because they maintain a more consistent flow than larger streams and are not as likely to flood and kill eggs and young. The high, turbulent water that follows rain or rapid snowmelt is much more common in the middle and lower reaches of a river, and is often destructive and sometimes calamitous to small fish and other organisms.

The trees that provide the basis for the life in a stream serve a number of important functions. They shade the water, keeping it cool enough to contain a great deal of dissolved oxygen. The pools produced by fallen trees give shelter to aquatic organisms during runoff following storms and spring melt. Fallen trees also provide critical cover for fish, protecting them from shore-ranging and flying predators.

*

In northern Michigan we have both marshy creeks and rocky creeks, originating either in boggy, spring-fed ponds or from surface springs seeping through the porous, sandy soil. I knew from studying topographical maps that the creek I followed that May afternoon began in springs. As I walked I imagined some dramatic origin, a bubbling pool cradled among cedars, with the creek gushing full-grown from it. I was only a little disappointed to discover that, like most streams, this one springs from the ground in a less dramatic way.

My shoes and pantlegs were soaked by the time I reached the source of the creek, if it can truly be called the source. The creek, I discovered, is formed by the confluence of a dozen minor tributaries, each dividing and dividing again until it becomes so small it is scarcely noticeable—lidded with leaning grass, slipping out of sight beneath the moss-covered humps of old trees in the swamp. At each confluence I took the strongest tributary, following first one, then another, then another until finally the one I chose had no more tributaries. It became too shallow to hide even a fingerling trout, barely deep enough to wet a shoe, and ended finally in a narrow, dark, aromatic gully among the cedars, where the ground was so wet I left a trail of slowly filling footprints. There in that tiny valley, with sides so close that trees leaned across and touched overhead, I found the springs that form the creek.

It begins with whispers and tears—gentle, nearly inaudible trickles seeping from the banks like wounds on a tree, each oozing enough water to nourish patches of swamp buttercups and wild peppermint and hummocks of moss thick as couch cushions. The seeps trickle downhill and gather at the bottom of the gully. It is there that the creek begins: six inches wide, half an inch deep, a rivulet trickling over rust-colored pebbles in a skinny bed lined both sides with moss. A hundred feet downstream the rivulet gathers the flow from a dozen other seeps and the water begins to fill with life.

Only big things get easily noticed. If we value just what is worthy of mention in newspapers and on television it is easy to believe that nothing much matters unless it is large enough to shake the earth. But standing in that gully crowded with growth and dampness, with newborn water leaking from the banks around me, I was reminded that great things often come from humble beginnings. In many ways it is the creek that makes the river.

7

MEANDERING RIVERS

Viewed from the sky, rivers are among the last nongeometric features of the landscape. Roads diagram the land in lines and right angles, towns and cities are made of squares and rectangles, farmers' fields are laid out in patterns of cubist artistry. Even many forests have squared-off borders.

Only rivers appear inviolable. They travel without plan, twisting like tendrils, as roundabout as idle conversation. Protected within corridors of vegetation, they can seem immune from encroachment—until you look closer and see a length of river that has been "channelized" for the convenience of barges, its bends ironed out like wrinkles. Then you see the sudden widening of a reservoir, the broad end blunted by the slash of a concrete dam, and all arguments of utility lose their force. A river interrupted is a travesty, a violation, an outrage. If you want to see people enraged, strangle the rivers they love.

Our love of rivers runs deep. Poets, painters, and composers have expressed it for thousands of years, for reasons most of us understand instinctively. While every star and snowflake and towheaded child is winding down to stillness, rivers win our hearts for seeming to defy universal laws. We are stirred by the perpetual flow of current, the effortless partnership

with gravity, the easily observed truth that you can never, as Heraclitus said, step into the same river twice. Like us, rivers spring from obscure sources and flow toward unavoidable destinations. If the sea represents eternity, the rivers that flow into it are the twisting, bold, and unstoppable currents of time.

It's hardly possible to speak of time without using the same words we use to describe moving water. Rivers and time both pass. They flow, sweep, run, hurry, and lag. They both change constantly yet never change. The connection between the two is fundamental and has inspired much comment:

"Who looks upon a river in a meditative hour, and is not reminded of the flux of all things?"

—Ralph Waldo Emerson, *Nature*

"Time is but the stream I go a-fishing in. I drink at it; but while I drink I see the sandy bottom and detect how shallow it is. Its thin current slides away, but eternity remains."

—Thoreau, *Walden*

"By the waters of life, by time, by time..."

—Thomas Wolfe, *Of Time and the River*

"To live by a river is to live by an image of Time."

—William Gass, "Mississippi"

"Time is the substance from which I am made. Time is a river which carries me along, but I am the river..."

—Jorge Luis Borges, *Labyrinths*

To travel by river you must adjust to river time. Follow that winding route and it becomes easy to believe it is a random, unpredictable course laid down by chance encounters, and that current and terrain prevent the river from getting on with the natural business of flowing straight and true.

It is only when you study a river from an elevation or examine it portrayed on a map that you recognize there is a pattern to its meandering.

Our word *meander* is derived from the old Roman name for Turkey's Menderes River, the *Maeander*, a waterway with a course so tortuous and winding that in ancient times it was thought it turned around and flowed backward. There is no strict definition of meandering, except that it is the sinuous pattern of river channels everywhere. It is a defining characteristic of rivers—a river with no bends, after all, is a canal—and is among the qualities we find most enchanting about them.

Their habit of meandering is one reason rivers are so often irresistible both for poets who celebrate nature's magic and scientists who explore its machinery. When conservationist Aldo Leopold turned his attention to meandering rivers he wrote, "For the last word in procrastination, go travel with a river reluctant to lose his freedom in the sea." His son, the renowned hydrologist Luna B. Leopold, studied the same phenomenon and wrote, "The typical meander shape is assumed because, in the absence of any other constraints, the sine-generated curve is the most probable path of a fixed length between two fixed points." With rivers, the magic and the machinery are so closely blended they cannot be separated.

Our appreciation of rivers originated, no doubt, in their usefulness. Rivers bring our freshwater to us and carry our wastes away. They water our fields and power our mills. They are highways to be traveled upon and fisheries to be plundered. Because they are so useful, they pass through history like the threads in a tapestry. The Amazon and the Mississippi, the Hudson and the Thames, the Yellow, Rhine, Missouri, Tigris, Orinoco, Rio Grande, Indus, Jordon, Ohio, Euphrates, Columbia, Congo, and Nile—rivers have shaped human history as surely as they have shaped the land. They have given birth to civilizations, powered the growth of cities, transported explorers and invaders. They have been the source of sustenance and the cause of sorrows. The greatness of the great rivers is often obscured—up close they too often reek of urine and fuel oil and are stained with the wastes of industry and agriculture—but their influence on human life has been immeasurable.

One small mark of their influence is that in many cultures rivers appear as subjects of mythology, often as sacred and purifying waters or as deities capable of inflicting floods, drownings, and other spiteful acts. To placate temperamental river gods, the Trojans tossed live horses into rapids and the Algonquin Indians threw tobacco into sacred waterfalls. In China, after the Yellow River flooded in 132 B.C., the emperor Wu Ti oversaw the repair of dikes by first sacrificing a white horse and a jade ring to the river.

In Hindu tradition, the Ganges River cleanses the sins of those who bathe in its waters and gives immortality to those who drown in it. Named for Ganga, the daughter of King Himalaya, the river was sent to earth because the gods took pity on the sinful state of humanity and offered people the chance to wash their sins away. The sacred river of Christianity and Judaism is the Jordan, the site of Jesus' baptism and a river long thought to be a branch of the stream that flowed beneath the Tree of Life in Eden. The Nile was central to the religion of the ancient Egyptians, just as it was central to the prosperity and security of their empire. The source of this great, mysterious river was thought to be Paradise. Its regular inundations, its habit of flowing highest during summer when other rivers were at their lowest, and its power to rejuvenate land that would otherwise be desert made it easy to credit the Nile with magical properties. It was believed, for instance, that barren women could be made fertile by drinking its muddy water, a notion that inspired a thriving trade in bottled Nile water between Egypt and western Europe as late as the Middle Ages.

If a river is worthy of worship, perhaps it is worthy also of punishment. The Greek historian Herodotus recounts the story of King Cyrus II of Persia, who led an army to attack Babylon in 539 B.C. Along the way, as the army prepared to cross the Gyndus River, one of the king's prized white horses bolted into the water, got caught in swift current, and drowned. Cyrus was so furious that he delayed his attack on Babylon for an entire summer in order to exact a singular revenge on the river. He ordered his soldiers to dig 360 canals radiating from the Gyndus, draining it in all directions. Only when the river was depleted of all its water did Cyrus and his army proceed to Babylon.

Efforts to control rivers are as old as human civilization. Excavations of 9,000-year-old settlements in Mesopotamia reveal complex irrigation systems connected to the Euphrates and Tigris rivers. The Egyptians built dams, canals, and artificial lakes to supplement the seasonal flooding of the Nile, and the ancient Chinese were so industrious in building canals between rivers that they compiled exhaustive encyclopedias devoted to them. In 20 B.C., the first of these "Waterway Classics" described canals and flood-control efforts on 137 rivers. By the sixth century A.D. the book had been enlarged to forty times its original length.

In more recent times, relentless efforts to contain and direct rivers have had mixed results. Diversion of water for irrigation makes it possible to raise crops in arid regions, but it turns many rivers into dry gulches during periods of scarcity. Power dams and their reservoirs produce a great deal of hydroelectricity, but in the process they flood large areas, forcing people from their homes and disrupting wildlife populations, and alter water temperature enough to affect downstream ecosystems. Dikes and levees meant to eliminate flooding on rivers like the Mississippi have raised river channels so high above their surrounding flood plains that when floods come—and they always do—the destruction is greater than ever before. In the 1960s, when the Army Corp of Engineers straightened 98 miles of Florida's Kissimmee River into a 56-mile canal, there was an immediate drop in the water level in the Everglades downstream and tens of thousands of acres of flood plain began to dry up. Waterfowl populations in the region declined ninety percent and three-quarters of the fish species native to the river disappeared. But since 1992 the Kissimmee's meanders have been restored. When completed in 2015, the Kissimmee River Restoration Project will have returned 44 miles of river and 20,000 acres of wetlands to its original condition.

Rivers give the illusion of progress, an inevitable and easily charted progression from origin to ending. From sufficient height and perspective we can see otherwise—a river that from the bank seems defined and permanent is actually temporary and easily directed. Even the water itself does not go anywhere, not in any final sense. That unstable liquid flows and tumbles downstream, mingles with the sea, disintegrates into component molecules that rise skyward, join with others, condense, fall, form rivulets, then creeks, then rivers again. Rivers are segments of a larger cycle but they are among the most visible and appealing of those segments, more accessible than ground water, more dependable than rain, more easily fathomed than oceans or lakes. Water is at its best when winding downhill between banks.

The Science of Meandering

It's in the nature of rivers to turn away from obstacles and seek the path of least resistance. But meandering differs from mere turning. It is consistent and regular and occurs regardless of whether there are obvious obstacles. In fact, it is most prevalent where the land is nearly flat, often in the lower portions of river valleys, where thousands of years of flooding and sedimentation have left thick layers of loose soil. Here, with few obstacles to deter it, a river should follow a straight line downhill. But it never does. Hydrologist Luna Leopold concluded that rivers almost never flow in straight lines longer than ten times their width. Why? What law does flowing water obey when it forsakes a direct path for one that winds?

The tendency to meander is consistent throughout the world. It can be seen in the smallest creeks and the largest rivers, in the currents of the Gulf Stream, in channels of meltwater running down the surface of a glacier, in trickles of rainwater streaming down the windshield of an automobile. During periods of low water in rivers dominated by riffles and pools, the deepest channels wander from side to side in a pattern identical to the paths of meandering rivers. In laboratories, water sent flowing down a straight channel in an inclined bed of fine sand begins deviating from its course almost immediately. It turns one way, then the other, vacillating in serpentine curves that at first are subtle but become steadily more pronounced. In just a few hours the river flows down an acutely meandering channel.

Most efforts to explain meandering begin with the inherent instability of water, which has been compared to the instability of a needle balanced on its point. In theory, the needle should remain balanced forever, but in practice it always falls over. Similarly, water always finds reasons to avoid running in a straight line. As a stream descends a slope it encounters countless forces that influence its flow and cause it to deviate. Some of the influences, such as large boulders and waterfalls, are obvious. Others, such as slight variations in terrain and the Coriolis force of the earth's rotation, are more subtle.

Once a flow deviates, even slightly, the deviation is magnified by centrifugal force. As a bend forms, current is faster on the outside, causing the bank to erode there. Water on the inside of the bank flows more slowly, allowing sediment to settle and accumulate in what is known as a point bar. The eroding power of the river is increased because surface water swept to the outside of the bend is forced downward in a scouring motion against the bank, then corkscrews along bottom to the inside of the bend, where the newly scoured soil is deposited. This combination of erosion on the outside

OXBOW

surface water

faster current

POINT BAR

sediment

AS OUTSIDE BANK ERODES
AND POINT BAR BUILDS,
MEANDER ENLARGES AND
MIGRATES LATERALLY.

of the bend and deposition on the inside causes the bend to gradually tighten. Gravity, meanwhile, causes the river to straighten briefly below the bend before being pulled into a curve in the opposite direction. Centrifugal force takes over on that bend as well and the process is repeated. The stream constantly seeks to correct itself, winding back and forth down the slope. In time it loops so radically that some of the meander bends become nearly circular and the river cuts new, more direct routes downstream. The peninsula around which the river once flowed now becomes an island. As the entrances to the bypassed channel fill with sediment, a crescent-shaped lake, called an oxbow—named for the U-shaped collar that fits around the neck of a harnessed ox—is left behind.

Because the surface of a valley slopes downstream, erosion occurs more quickly on the downstream side of each bend in the river. As a result, the meanders migrate slowly downhill. How fast a river migrates in this way depends on its volume and velocity and the ground it flows over. Rivers in soft, easily eroded soil carry a lot of sediment, which acts like the sand in a sandblaster to chew up objects in its path, speeding the migration. In time a meandering river passes down the entire width and length of a valley, leaving a record of its passage in oxbows and abandoned channels.

Hydrologists have concluded that meanders occur because they save work for rivers. Flowing water seeks consistency. It tries to make shallow parts deeper and deep parts shallower and to smooth out waterfalls, rapids, and other rough spots. In computer models of river flow, the most efficient flow shape for equalizing all the variables a river is likely to encounter is the meander. It allows a turning river to do the least amount of work as it turns, and helps distribute energy equally from bend to bend. Probability theory predicts that all rivers eventually "discover" that a meandering pattern is the most efficient use of its energy. As in so many of nature's designs, an efficiently meandering river happens to form a pattern we find visually appealing.

*

Our language makers must have lived near rivers. We think in streams of consciousness and speak in sweeping channels of utterance. If we speak well we're told we are "fluent," and no one is surprised to learn that the word is derived from the Latin "to flow" and shares roots with *fluid* and *flux* and *effluence*. Listen carefully to flowing water and you can hear babbling, murmuring, and whispering sounds so like human voices that you wonder if rivers were the primal model for the sounds we make. In writing we shape our thoughts into sentences and string them together until they look like a river meandering across a plain. The words line back and forth down the page, slowing at commas, eddying between parentheses (a brief loitering, a small backwater of meaning), gathering force and momentum until they spill off the bottom of the page like water falling off a continent into an ocean.

In an Inuit legend all rivers once flowed both uphill and down, but Raven, the creator of the world, decided that such a convenient arrangement made life too easy for people. We have been riding the current or fighting it ever since. Those who ride it are river people. River people know that current can be resisted but never defeated. They abandon themselves to flow and flux, accept the value of boundaries and the power of persistent current, understand that loved ones, accomplishments, opinions, possessions, and all the carefully fashioned handiwork of a lifetime fall eventually to the surface and are carried away downstream. If you are a river person you can never cross a bridge without looking for a glimpse of water. You remember your first river trip in a canoe and the intoxicating sensation that you were getting something for nothing, that you were stealing a free ride, a gratis trip on the merry-go-round. You remember the dizzying conviction that you and the river were stationary, while the land unreeled beside you in a continuous play of background and perspective.

Rivers enchant us and somehow pry into the nebulous pleasure center in our souls. I think it is a matter, largely, of earning the right to be enchanted. We care most deeply about rivers when we have invested a great deal of effort in working on them or have spent much time boating, fishing, exploring, or observing them. As in any love affair, you get back only as much as you put in. It's a tribute to the value we place on moving water that so many people are willing to put so much into the affair.

8

THE ANATOMY
OF RIVERS

A friend and I once spent most of our free time for two years canoeing and fishing the lengths of nearly fifty rivers. Now, a decade later, I'm certain that we could be blindfolded and transported to any of those rivers and know immediately where we were. Every river is unique, with unmistakable qualities of aroma, color and clarity of water, and types of bottom, banks, and surrounding topography. Recognizing a river after being away from it for a while can be as easy as spotting an old friend in a crowded room.

Every river is unique but certain features are nearly universal. On most rivers gradient tends to be steep in the upper reaches and decrease as the river descends. Rivers with gravel beds tend to alternate pools and riffles, and, for reasons that are unknown, their riffles tend to be spaced five to seven river-widths apart. Most rivers widen and deepen and their volume increases as they proceed downstream—unless of course their water is diverted for irrigation or other purposes, or enters an arid region that

absorbs and evaporates it. River valleys are narrow at their tops and widen toward their lower ends. The number of tributaries flowing into a river decreases in a mathematical progression as you proceed downstream.

The Scottish geologist John Playfair was so intent on pinning down a few certain things about rivers that in 1802 he declared (in a sentence that reads not unlike a tightly meandering river) what has since become known as Playfair's law: "Every river appears to consist of a main trunk, fed from a variety of branches, each running in a valley proportional to its size, and all of them together forming a system of valleys connecting with one another, and having such a nice adjustment of their declivities that none of them join the principal valley at either too high or too low a level; a circumstance which would be infinitely improbable if each of these valleys were not the work of the stream which flows in it."

I love to follow rivers just to see the nice adjustment of their declivities. It's one of their best features. Another way of stating Playfair's law might be: You can't separate a river from its valley. The valley is shaped by the river and the river is shaped by the valley. They direct and color each other's future and hold up geologic mirrors that reveal each other's past. This was a revolutionary concept in Playfair's day, when almost everyone assumed that valleys were God's invention and rivers flowed through them merely because they offered the course of least resistance.

From the air we can see just how dramatically the surface of the earth has been shaped by water. The dominant feature of the moon is craters caused by the impact of meteorites and comets. On earth the dominant feature of the land is river valleys. Flowing water is a tireless sculptor. It carves valleys and canyons, divides forests, traces serpentine furrows through plains, and builds broad deltas on ocean shores.

An aerial view also reveals several distinct types of river valleys, each determined by geologic vagaries. Most rivers and their tributaries, when allowed to flow without unusual restrictions, create a pattern that is dendritic, similar to the branching of trees and the veining of leaves, with distinctive Y-shaped junctions where tributaries join together.

In regions of steep valleys and ridges, where geologic faults can force water to turn at abrupt angles (like the Delaware Water Gap and Kittatinny Ridge in New Jersey and Pennsylvania), streams sometimes flow in parallel valleys fed by tributaries entering at right angles, creating trellised river systems. On plateaus or where elevations are formed from the dome-shaped remains of volcanoes such as Mount Hood in Oregon and the Black Hills of South Dakota, streams near the tops of mountains and hills flow outward over the flanks in a radial pattern. Pinnate (or feathered) river systems are similar to dendritic systems except that the primary streams all flow to one side and their tributaries enter at an oblique angle, like the barbs on a feather. Parallel drainage patterns occur where the gradient is steep and uniform, allowing both main streams and tributaries to flow directly downhill, parallel to one another. Rectangular patterns result when rivers and their tributaries tend to flow in only two directions, describing right angles when they join. Anastomosing (or intercommunicating) river systems are found in tidal marshes, plains, and deltas, where meandering channels create many oxbow lakes, abandoned courses, and dead-end channels. Braiding those features creates a complex system of interconnected channels, both wide and narrow, deep and shallow.

Rivers can have several possible sources. Most are the result of springs and small streams coming together. Some, like the Mississippi and the Nile, originate in lakes. Others, such as Europe's Rhine and the Rio Grande in Mexico and the United States, gather melting snow and ice at their sources.

Regardless of how they begin, each river is composed of a collecting system that funnels water through its tributaries to the main stream, which in turn channels water to an ocean, sea, lake, or dry basin. Within those channels the flowing water is always composed of two kinds of flow. The bulk of it proceeds downstream in *turbulent flow*—the irregular swirls, surges, and eddies that are such fascinating ingredients of moving water. When colored dyes are released into rivers, turbulent flow is revealed to be erratic and unpredictable, traveling downward, upward, and sideways as it makes general progress downstream.

Laminar flow is less easy to notice. It runs in straight lines, with much less velocity than turbulent flow, in a thin layer against the bottom and banks of a stream. Water is very susceptible to friction. It flows easily over a smooth, lubricated surface, but is slowed when it passes over rough, uneven surfaces. Where water rubs against the sand or stone of a streambed or the rocks and roots along a bank, friction slows it and creates the narrow layers of laminar flow. Usually the only way to notice those bands of slower water is to watch bits of sand and organic matter in the stream. Such particles are swept quickly away when they are released on the surface or at the middle depths of a stream, but when they're released on the bottom they linger. Grains of sand that weigh almost nothing will dance and bob there as if tethered to the foundation of the river.

THE VELOCITY OF A RIVER

The fastest water in a river is always found where there is the least friction: just beneath the surface in the deepest water near the center of the channel. As a river bends, that fastest water swings to the outside of the bend, leaving a zone of slower water on the inside.

In a boat you can feel the difference immediately when you leave the deep, fast water on the outside and enter the shallower, slower water along the inside of a bend. Such differences in flow make it possible to paddle or pole upstream against the current in even surprisingly fast rivers if you follow the edges and take advantage of eddies. Trying to force your way against a midstream current is strenuous exercise but a poor way to make progress.

Typically, rivers with the most velocity are those with channels shaped, in cross section, like a semicircle. Rivers that are wide and shallow or deep and narrow tend to have a slower flow because their water is slowed by friction against the greater surface area of their beds and banks. For the same reason, engineers have long known that water can be transported most efficiently in pipes that are circular rather than oval, square, triangular, or trapezoidal. A circle offers the least possible surface area for the volume it contains and thus creates the least friction. An interesting characteristic of rivers is that they always seek their ideal shape. A deep, narrow river tends to erode its banks outward, and a wide, shallow river tends to erode its bed deeper. In both cases, the tendency is to adjust the channel's shape until it becomes a half-circle and reaches equilibrium between the flowing water and the area it flows through.

Reaching the ideal shape is seldom possible, however, because other forces act upon it. When a river passes through a region of very loose, easily shifted sand or glacial debris, the river transports more bed material than can be disposed. If the banks are weak the river will spread out, creating a wide, shallow bed and often forming *braided channels* separated by small temporary islands. Such channels are common in glacial rivers, where a great deal of sediment produced by glacial scouring is washed down the river and settles in broad outwash plains. Some of the braided channels in rivers in Alaska, Iceland, New Zealand, and elsewhere spread a half-mile or more in width, with water only a foot or two deep. Braided rivers are often lacking in living organisms because they shift too frequently for populations to become established and because a bottom of fine sand or sediments offers so little cover. Islands in a braided river come and go rapidly unless anchored by vegetation, and the channel never reaches a half-circle shape because as quickly as loose bed material is flushed away more comes from upstream to replace it.

The velocity of a river is proportional to the gradient of its channel. That's obvious. Anyone can see that the fastest rivers are the steepest ones.

But if you have canoed or kayaked in powerful rivers you know that the speed and power of current vary a great deal from place to place, and not just because of gradient. The current's velocity increases when it passes through a chute between rocks, for instance, then slows nearly to a stop in the eddy behind the rocks. In even the most furious rivers there are always eddies where the current is slowed by obstructions. A paddler who learns to recognize those spots can "eddy-out" in them, turning a boat upstream and coming to rest there. Anglers know that such eddies are good places to find fish.

A river's velocity also varies with its volume. The greater the volume, the faster the river flows. Rapids are usually formed in a river when it descends at a rate of at least fifteen feet for every mile of length. Rivers with high water volume—the Colorado comes to mind—are plenty challenging with that amount of descent, while similar gradient on a small river with little volume is likely to create insignificant riffles. A small river might descend thirty or forty feet in a mile before it produces the "backrollers," "haystacks," and other features that attract whitewater paddlers.

Less obvious is the effect of water temperature on river velocity. Like motor oil, water increases in viscosity as its temperature drops. Viscosity is the quality of being glutinous in consistency, of being sticky, thick, or adhesive. The hydrogen bonds connecting molecules of water are affected by temperature and become stronger as the temperature decreases (causing water to reach its maximum density at about 39 degrees Fahrenheit). Cold water is dense and sticky. When it encounters a rock or other obstacle in a moving stream, it is slowed more than when it is warm. As the water's temperature increases and the viscosity decreases, the water flows a little faster. Water above 39 degrees and below 68 degrees Fahrenheit increases its velocity about one-half percent for every one-degree increase in temperature. Silt and other particulate matter sinks about twice as fast when the water is 73 degrees Fahrenheit than when it is 32 degrees. Thus a warm river is faster and carries less silt than a cold one.

The measure of a river's volume and velocity is its *discharge*: the amount of water that passes a specific point in a specific period of time, measured in cubic feet per second (cfs). Small streams can have a discharge of only a few cubic feet per second, while the Mississippi's discharge ranges seasonally from as low as 1.5 million to as high as 12 million cfs. Such extremes are common on many rivers. The Colorado River through the Grand Canyon normally fluctuates between 4,000 and 90,000 cfs, but in 1921 a flood raised the flow to more than 200,000 cfs. The Amazon has the greatest discharge of all rivers: It has been recorded as high as 52.5 million cfs.

READING A RIVER

Canoeists, kayakers, anglers, and river pilots speak of being able to "read" a river. With experience, they learn to decipher features on the surface as easily as they decipher words on a page, a comparison noted by Mark

Twain in *Life on the Mississippi*: "The face of the water, in time, became a wonderful book—a book that was a dead language to the uneducated passenger, but which told its mind to me without reserve, delivering its most cherished secrets as clearly as if it uttered them with a voice. And it was not a book to be read once and thrown aside, for it had a new story to tell every day."

The stories change as the river changes, but the key elements are always the same. If a river flows quickly over an uneven bottom, there will be rapids, that utterly appropriate word for the turbulence that results when water flowing with high velocity meets obstacles. One way the turbulence reveals itself is with a churning mixture of water and air, or white water. When the water passes over a large boulder, ledge, or other obstacle, it humps above the obstacle, plunging over it and creating an eddy or souse hole or back-curling wave (or all three) immediately downstream. Farther downstream the turbulence is likely to be visible in a series of standing waves. Standing waves are motion made stationary, energy organized. If they are large they are known by paddlers as "haystacks" for the way their crests break in a perpetual upstream curl. They remain in place while the water passes through and beyond—just the opposite of waves in a lake or ocean. Whitewater paddlers quickly discover that waves standing in a rapids can sustain a surfing kayak indefinitely or can toss aside a raft loaded with half a ton of passengers as easily as if it were a chip of balsa.

Most rivers with rapids are at least somewhat seasonal, with high water during spring or summer producing immense, spectacular rapids that other times of the year might be reduced to riffles. One clue to such seasonal velocity can be found in the relative size of stones in a riverbed. River flow causes stones to shift downstream, sorting them by size at the point where discharge is no longer sufficient to transport them. The upper reaches of large, powerful rivers like those in the Pacific Northwest of North America, for instance, are sometimes paved with Volkswagen-sized boulders, everything smaller having long ago been sent tumbling away by spring floods. Miles downstream the shores and bottom might be cobbled with smaller rocks strangely uniform in size, say the size of cantaloupes and pumpkins. Downstream farther yet are nothing but stones the size of baseballs; below that, gravel and sand and, ultimately, silt and mud. If you find such a distinct sorting of stones on a river you can be sure that during some seasons at least, that river roars.

Charles Darwin witnessed this sorting process at work on the Maypu River in Chile and described it vividly in *The Voyage of the Beagle*:

> Amidst the din of rushing waters, the noise from the stones, as they
> rattled one over another, was most distinctly audible even from
> a distance. This rattling noise, night and day, may be heard along
> the whole course of the torrent. The sound spoke eloquently to the
> geologist; the thousands and thousands of stones, which, striking
> against each other, made the one dull uniform sound, were all
> hurrying in one direction. It was like thinking on time, where the
> minute that now glides past is irrecoverable. So was it with these
> stones; the ocean is theireternity, and each note of that wild music told
> of one more step towards their destiny.

"In time and with water, everything changes," wrote Leonardo da Vinci. The truth of his observation is everywhere apparent on our planet, from the great river valley of the Rhone to the dry gulches of the Dakota Badlands to the broad fanning delta of the Mississippi. It is especially apparent where rivers flow, their handiwork the result of many years of work. The upper Colorado River is more than 25 million years old. It takes time on a grand scale to carve Grand Canyons.

Water's work can be seen on a smaller scale as well. Where a rapids flows over stationary bedrock, it is often possible to find conical holes in the rock so uniform they appear to have been made by human hands. These "potholes" occur when small whirlpools of water carrying sand and small stones drill patiently into the rock. Potholes can be as small as a few inches or as large as a few feet in diameter and are often several feet deep. Some rapids create so many of them that they overlap and drill through the rock to form gorges, their scoured and rounded edges visible all up and down the gorge walls. They can be carved in surprisingly little time. In one case, potholes were observed forming 10 feet deep in limestone after only 18 months. On another river, where the bed was granite, potholes 5 feet in diameter took 75 years to form.

The dynamic nature of rivers is most apparent when you get to know a river well enough to see its changes year after year. My own favorite river is small and quick-spirited, a spring-fed riffle-and-pool trout stream hurrying through a valley of mixed hardwoods, aspens, and conifers in the hills of northern Michigan. It flows clean, clear, and cold over a bottom of gravel and sand, beneath the branches of leaning cedars known locally as "sweepers," past sandy washouts and ancient silvered stumps that are remnants of log drives more than a century old. In the evenings, when the wind falls and the rich scent of swamp rises to flow down the valley like a river above the river, you can go there to cast dry flies for brook trout and brown trout or just to wade into the water and feel the insistent nudge of the current. If you stand in midstream and lower yourself until you are eye level with the surface and look upstream, you can watch the river slide downhill toward you. The descent is distinctive and measurable, perhaps a foot for every 250 feet. Watch it long enough and the whole broad face of the river seems to fall beneath you. If you listen you can hear the gentle clatter and scrape of the earth being changed.

STREAM CAPTURE

Erosion causes riverbeds to gradually migrate upstream, a tendency that sometimes results in competition between river systems. A powerful river erodes its valley more quickly and deeply than weaker neighboring rivers. If it erodes all the way through a hill or mountain that divides watersheds it can engulf the top of a neighboring system, intercepting springs, creeks, marshes, and other headwater sources and causing them to flow into the new valley, cutting off the water supply of the weaker river.

Such piracy can have several consequences. The river system that captures a greater drainage area gains more discharge of water, while the "beheaded"

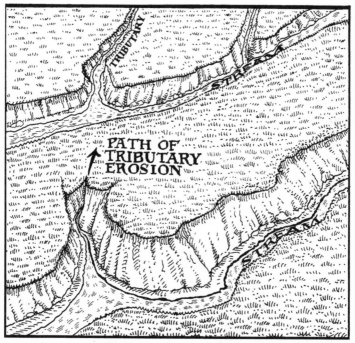

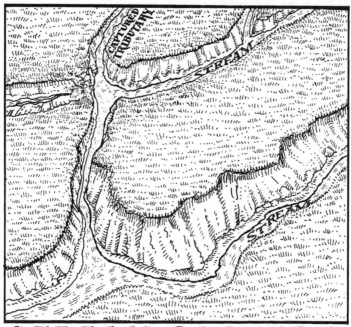

·STREAM CAPTURE·

system is depleted, leaving behind gorges that lead nowhere and valleys where water no longer flows. Fish and other aquatic wildlife native to one watershed can be spread to others, resulting in greater distribution of some species and possible extinction of those that cannot compete with newly arrived species.

Stream capture is common in many places. Rivers in the Appalachian Mountains cut upstream wherever the rock is easily eroded and sever the tributaries of lesser rivers. For thousands of years the Pecos River in Texas and New Mexico has been eroding steadily upstream and north, capturing a series of now-defunct rivers that flowed east from the Rocky Mountains across the Great Plains. The water that once ran across the Texas High Plains now flows down the Pecos to the Rio Grande.

EPHEMERAL AND INTERMITTENT STREAMS

Rivers get their water from either continuous or occasional sources. The continuous sources are usually lakes or groundwater seepage. The occasional sources are precipitation or melting snow and ice. If the continuous sources are reliable enough, the river can maintain a steady flow year-round, surging to higher levels when rain or snowmelt swells the flow. If the river is fed by no continuous sources, however—its water coming only from snow or rain—then it will lose its water entirely at certain times of the year. Such a river is *ephemeral* if it flows only with runoff after rainstorms. It is *intermittent* if it flows both after rain and during wet seasons, when it is fed by recharged water tables.

In the Sahara, the beds of ephemeral rivers are known as wadies and are so effectively smoothed by past floods that when dry they are often used as roads. On occasion caravans following wadies have been met by sudden, downrushing floods caused by rainstorms dozens of miles away. The same hazard exists in the arid southwest of North America, where ephemeral rivers flow down *arroyos* (Spanish for creek or gulch). Campers who choose an arroyo as a campsite because it is convenient, level, and clean risk being washed away by a flash flood as they sleep.

Temporary streams and ponds can be home to a number of specially adapted animals that have learned to get along when the water disappears. Some merely retreat ahead of the shrinking water, either to deeper water downstream or to other streams. Some fish, crustaceans, snails, and flatworms take refuge in pools left behind when the shallower channels have dried up and have adapted to living in water with high temperature and low oxygen content.

A number of crayfish, flatworms, snails, mites, and beetles can survive dry seasons by burrowing as deep as several feet into the bottom

of drying streams until they reach water. Some ensure survival of their species by producing eggs that can survive long periods without water, and depositing them early in the season when water is abundant, before the droughts of summer.

Some inhabitants of temporary streams are able to survive even when the water disappears entirely. Among them are snails that seal their openings with a dry, impervious mucus, and caddisflies whose larvae build cases tight enough to prevent drying out. Four species of lungfish in Africa and one each in South America and Australia can survive in rivers, ponds, and swamps that dry up during the dry season by burrowing into wet mud when most of the water is gone, curling into a ball, and secreting a lining of protective mucus. As the mud dries around the fish the mucus becomes a parchment that traps the moisture in the fish's body, preventing desiccation. A pair of lungs inside the gut of the fish allows it to breathe air through its mouth. The air comes via a narrow opening to the surface left in the mud as the fish burrowed down, and is drawn underground by the pumping motion of the fish's throat muscles. As in any lung, the pouches are lined with blood vessels that absorb oxygen from the air. Lungfish can survive many months waiting for the waters to rise again. When they do, the fish breaks free of its muddy shelter and breathes again through its gills—unless the water is oxygen-depleted, in which case it rises to the surface and gulps air.

Although more often found in temporary ponds than intermittent streams, the water-holding frogs of central Australia, members of the subfamily Cycloraninae, demonstrate a tenacity equal to the lungfishes' in the face of hardship. The frogs appear during rainy periods that may occur only every two or three years. While the wet season lasts—rarely longer than a few days or a week—the frogs feed, mate, and lay eggs in shallow pools. The eggs hatch and the tadpoles mature, all within those few days. As the water evaporates and drains away, each frog, young and adult alike, absorbs so much water it becomes bloated until it is nearly spherical, then burrows deeply into the sand and excavates a small chamber for itself. Once the chamber is complete, the frog secretes a membrane that covers its body surface to prevent evaporation, leaving only tiny openings over the nostrils to allow breathing. It can remain alive this way for at least two years, waiting patiently for the next rain to rouse it from its sleep.

LIFE BENEATH A RIVER

We tend to think of a river existing in a cleanly defined zone bounded by the banks beside it and the bed on its bottom. But a river flows as well through all the spaces between rocks, gravel, sand, and silt many feet beneath and to the sides of the visibly flowing water. This basement of a river is known as the *hyporheic zone* (from the Greek "to flow beneath"). It varies in depth according to the composition of the river's bed and the size of the river, but even where the bottom is packed with the smallest particles the zone extends at least a few inches deep. More typically it extends for several feet. In those few inches to a few feet beneath the river lives a unique and remarkable community.

Oxygen is a limiting factor in how much life the hyporheic zone can support, with cold rivers at high altitudes supporting the most life. Oxygen disappears quickly in fine sediment, so the layer of habitable zone is narrow. But where sediments are coarse the movement of water can continue for a considerable depth, blending the moving water of the river with the much

slower groundwater beneath it. It might move only an inch or less each hour, and the oxygen content is likely to be low and the carbon dioxide level high, yet it is an attractive habitat for a surprising variety of flatworms, Oligochaeta (segmented worms), tardigrades, copepods, mites, water fleas, scuds, stoneflies, caddisflies, and mayflies. Many of those animals live also on the bottom of the river; some are found nowhere else. Researchers have noted that the organisms vary the depth at which they live, moving deeper during the winter and when the rivers are in flood. The total numbers of creatures involved can be enormous. One researcher counted more than 34,000 organisms per square meter in the four inches beneath a stream in Germany. Two Canadian biologists consistently found more invertebrates living in gravel two to six inches under an Ontario stream than on the bottom itself.

9

THE BIRD IN THE
WATERFALL

Arturo Redlich has witnessed every kind of misbehavior and calamity during the three decades he's been guiding trout fishermen in southern Chile, so he was not surprised that April morning when I fell off the cliff above the rapids of the Rio Puelo.

Me? I was surprised. I blamed the fall on a plump gray songbird I once watched swimming beneath the surface of a fast Wyoming river. The bird, an American dipper, alerted me to the fact that whitewater rivers are not only a pleasure to look at and listen to, but are home to great numbers of unusual birds, fish, insects, and other animals. I would never again see a rapids without wondering what lived in that rollicking water.

Which was just what I was wondering that day on the Rio Puelo, with the Andes looming overhead and the river roaring through a gorge that looked like somebody had parted the mountains with a crowbar. You have to understand, the Rio Puelo is drop-dead gorgeous: pristine and violent, colored a brilliant, otherworldly blue, the kind of river that makes your throat clench and your heart rip. It's home to large numbers of ridiculously

Cinclus
mexicanus

AMERICAN
DIPPER

TORRENT DUCK
Merganetta armata

big rainbow and brown trout, descendents of fingerlings transported across the Atlantic in the nineteenth and early twentieth centuries. It's home also to an uncommon species of waterfowl that doesn't look anything like the American dipper but has similar tastes. The torrent duck, *Merganetta armata*, is slim and streamlined, with powerful legs and broad webbed feet, a stiff tail that works as a rudder while the duck swims underwater, and claws sharp enough to give it purchase on slippery rocks. It lives, as its name suggests, in torrential rivers where it feeds by swimming to the bottom and preying on aquatic insect larvae, crustaceans, and small fish. Only a few torrent-duck nests have been found, but all have been located among rocks bordering rapids or on cliffs above them. Torrent ducklings are thought to leave their nests within an hour of hatching and are immediately able to plunge 60 feet from a cliff into the river below and swim fearlessly across— and dive beneath—even the wildest rapids.

When Arturo learned of my interest in the torrent duck he told me he knew a rapids where a pair of them could sometimes be seen. To reach the rapids we motored five miles in a Zodiac raft across a lake surrounded by snow-capped mountains. With us were a pair of young Chilean women catching a ride downstream to the village at the mouth of the river, where they planned to spend the day with friends. They were lovely, shy, graceful women of perhaps 18, dressed in tight jeans and leather jackets and wearing enormous dangling earrings. They were so tiny that they made me feel grotesquely large, like a character in a medieval painting by an artist who had not yet figured out the secrets of perspective.

At the end of the lake a gap in the canyon walls allowed water to escape in a rush downstream toward the Pacific. Arturo slowed the engine to an idle when we approached the outlet. Below us we could see the rapids. They consisted of a series of short, heavy, standing waves breaking with the ponderous weight of rapids in the Grand Canyon section of the Colorado. They were not particularly fierce—I've canoed in water nearly as fast and in waves nearly as formidable—but they were *big*. They were a rapids with volume, capable of swallowing houseboats. Any creature that could swim in such water had my respect.

We went to shore at the top of the rapids and docked the raft. A crude wooden ladder had been mounted against the bank to allow an eight-foot climb from the water to a farmer's goat pasture above. I followed the women up the ladder to the top and turned to scan the rapids. Right away I saw a torrent duck. Or I think I did. Something fast and dark skipped across the waves and threw up a trail of spray. It disappeared in a sudden flip on the far side of the river.

"There!" I shouted. "Arturo! Is that a torrent duck?"

I took a step forward to point toward the spot where the duck disappeared and my foot slipped on the edge of the bank and I fell. I would have fallen all the way to the water except that I caught myself halfway down in a tangle of exposed roots and hung there like a fly in a spider's web. The Chilean girls looked at each other. One actually rolled her eyes. I'd been fancying myself the sophisticated norteamericano but clearly they saw me for what I really am.

"El gringo es poco loco," Arturo said from the raft below.

The young ladies looked solemnly at the ground and whispered, "Sí."

*

Fast rivers are distracting enough, even if all you watch is the water. They're beautiful and enchanting, so filled with color, light, movement, and music that they can hold your attention for hours. Watch them long enough and you want to draw closer. You want to become intimate, to embrace them, to immerse yourself in them, which is no doubt why so many of us enter them in waders, canoes, kayaks, and inflated rafts.

Wildlife choose to live in fast water for more utilitarian reasons. Whitewater is white because it is saturated with air. Oxygen dissolves easily in water, making rapids attractive to organisms willing to put up with the inconvenience of racing current in exchange for the oxygen bounty. Prominent among those creatures are aquatic insects, crustaceans, and mollusks, which feed by grazing on vegetable matter, filtering small organisms from the water, or preying on one another. Their presence attracts predators, including trout and other fish that have the strength and body shape to maneuver in fast water. Phoebes, flycatchers, swallows, and other songbirds range above the rivers or along their edges feeding on insects as they emerge from the surface or swim in the shallows. A few waterfowl, like the blue duck of New Zealand, the Salvadori duck of New Guinea, the harlequin duck of the Arctic and subarctic, and the African black duck, can sometimes be found feeding on aquatic insects and small fish in river rapids. But only a few birds prefer the fastest water. The torrent duck of the Andes is notable for that preference. So are the dippers, which are the only songbirds adapted for swimming and walking underwater in fast-flowing, cold-running waters.

The five species of dippers that make up the family Cinclidae are found on four continents. They include the North American dipper (*Cinclus mexicanus*); the white-throated dipper of India and Pakistan (*C. cinclus*); the brown dipper of Africa and Indochina (*C. pallasii*); and South America's white-capped dipper (*C. leucocephalus*) and rufous-throated dipper (*C. schulzi*). All are similar in size and appearance and all live in similar habitats.

The American dipper is present in fairly reliable numbers, about one per mile, on most of the fast mountain rivers in the Rockies and other mountains of western North America. It shares the water with trout, feeding on the same aquatic insects and small minnows that are the primary prey of cutthroats, browns, brookies, and rainbows. The dipper is distinguished for having adapted to a semi-aquatic life without any obvious physical advantages for it. It is squat and plump and short-tailed, with feet that are slightly oversized but otherwise typical of the passerines, or perching birds. Other than a nictitating membrane over each eye, scales that seal the nostrils while underwater, and a large oil gland at the base of the tail used to waterproof its feathers, the dipper appears to be better suited to suburban lawns than mountain torrents.

Yet it lives successfully year-round among the waterfalls and whitewater rapids of streams and rivers in mountains from the Pacific coast to the eastern slope of the Rockies, from northern Alaska to southern Mexico. Part of its success in living in harsh environments is due to a layer of down next to the skin that serves as such effective insulation that on warm days the bird must stand in cold water to keep from overheating. During the winter the dipper has earned a reputation as a tough guy by singing cheerfully during blizzards and swimming nonchalantly in water rimmed with ice.

The nature writer John Muir was an eloquent and passionate champion of the American dipper, or water ouzel, as he preferred to call it. His description of the bird in *The Mountains of California*, published in 1894, still stands among the most vivid ever written:

> The waterfalls of the Sierra are frequented by only one bird, — the Ouzel or Water Thrush. He is a singularly joyous and lovable little fellow, about the size of a robin, clad in a plain waterproof suit of bluish gray, with a tinge of chocolate on the head and shoulders. In form he is about as smoothly plump and compact as a pebble that has been whirled in a pot-hole, the flowing contour of his body being interrupted only by his strong feet and bill, the crisp wing-tips, and the up-slanted wren-like tail.

Muir noted that he often saw the birds singing cheerfully in the worst imaginable weather. As long as there was open water, there were dippers plunging into the tumbling, ice-cold rivers of Yosemite or diving as deep as 30 feet beneath the surface of frigid alpine lakes. Wrote Muir,

> He is the mountain streams' own darling, the humming-bird of blooming waters, loving rocky ripple-slopes and sheets of foam as a bee loves flowers, as a lark loves sunshine and meadows. Among all the mountain birds, none has cheered me so much in my lonely wanderings, — none so unfailingly. For both in winter and summer he sings, sweetly, cheerily, independent alike of sunshine and of love, requiring no other inspiration than the stream on which he dwells. While water sings, so must he, in heat or cold, calm or storm, ever attuning his voice in sure accord; low in the drought of summer and the drought of winter, but never silent.

Though still sometimes referred to as the ouzel, water ouzel, water thrush, or teeter bird, the most widely accepted name for *Cinclus mexicanus* is the dipper. The common name refers to the habitual bobbing motion of the bird, especially when it is young, as it raises and lowers itself vertically an inch or so by bending its legs into a crouch then lifting to a standing position. It does this repeatedly, frequently, and rapidly, sometimes 40 to 60 times per minute. The dipping behavior could be a form of communication used by the birds to locate one another along noisy streams, where calls might not be heard. Or it might be a vestige of behavior from the nest, where hungry nestlings raise and lower themselves frantically to attract the attention of their parents.

The dipper's nictitating membrane, or inner eyelid, is often assumed to cover the eye and protect it while the bird is underwater, but some ornithologists have challenged that assumption, pointing out that the membrane is translucent, not transparent, and thus probably inhibits vision underwater when the bird is most in need of clear sight. One suggestion (based on analysis of films made of dippers while blinking) is that the nictitating membrane is used to clear water from the dipper's eyes while it is nesting, roosting, or resting beside waterfalls and rapids where spray would otherwise obscure vision. When the dipper blinks, the membrane is revealed in a quick white flash that is possibly a signal of alarm or aggression.

Few birds have given themselves as completely to their environment than the dipper. It is so focused on rivers and streams that it seldom flies far from the water, preferring to follow river courses faithfully rather than take even a short bypass overland. The dipper feeds, rests, courts, and mates beside or in running water. It builds its nest of moss in a camouflaged roofed dome at the virtual edge of the water—on the underside of a bridge, the top of a midstream rock, or, frequently, beside or behind a waterfall. The nest is often so near falls and whitewater that a constant dousing with spray keeps the moss it is built with alive and growing. From the moment young dippers hatch they are steeped in the sounds, scents, and touch of rapidly flowing water. They are somewhat more precocial than most passerines, fledged and able to leave the nest about 18 days after hatching, and immediately able to run, climb, and flutter along rocks and swim and dive in fast water.

The dipper I watched years ago in Wyoming spent its days hunting for food along a single stretch of river. I fished the same stretch nearly every day that September, walking the edges of water too deep and powerful to wade across, picking my way over the slick cobblestone bottom, casting across a surface jumping with white-capped waves. My imitation stonefly nymphs were taken with heartening regularity by trout 14 or 16 inches long. While I fished, the dipper hopped among the rocks at the fringe of the river, exploring with all the single-minded absorption of a dog investigating shrubbery for odors. We kept ourselves busy, neither minding the other's presence, both occupied with insects and their larvae—I casting counterfeits, he catching and eating the real thing. Occasionally the bird flitted past me, his wings nearly touching the water, or half-flew and half-hopped across rocks behind me. A few times I watched him plunge beneath the surface, disappearing for seconds, then popping to the surface and jumping to the top of a rock, where he gave a brisk shake to dry his feathers. Once he scooted into the river in front of me and took a stroll along the bottom with his wings held out for balance. When the current caught him, he turned and soared away underwater.

Muir entertained the idea that the dipper's fluid song, its bobbing motion, and its seemingly cheerful demeanor were reflections of its environment: "Ouzels seem so completely part and parcel of the streams they inhabit, they scarce suggest any other origin than the streams themselves; and one might almost be pardoned in fancying they come directly from the living waters, like flowers from the ground." Such a romantic idea has intriguing implications. A bird that lives thoroughly connected to a river is bound to be a river bird in every sense. If, like dippers, we come to resemble the places we inhabit, we have good reasons to spend as much time as possible near rivers.

In a rapids things happen fast and are easily missed. So it's appropriate that my only view of a torrent duck should have been a glancing one, a not-very-certain glimpse of a sprinting, splashing streak across the surface.

Maybe someday I'll go back to Chile and get a better look. And maybe someday along a trout stream in Wyoming or Oregon I'll see a dipper emerge from its next behind a waterfall, erupting in a burst of luminous spray as if water and air were giving birth to a bird.

LIFE IN THE FAST LANE

Like the creatures that live in another challenging environment washed by powerful currents—the edges of the oceans—animals in fast rivers must hold themselves in position or be swept away. To help them keep a grip on the world they have evolved a dazzling variety of hooks, claws, grapples, suctions, cement-forming glands, and other hold-fast devices.

Larvae of widely distributed aquatic insects such as mayflies, stoneflies, and caddisflies use remarkably similar devices and tactics for clinging to the bottom. These "torrential fauna" often have flattened bodies that allow them to keep a low profile while crawling about in the boundary layer at the bottom of a river. Many others are streamlined with "fusiform" bodies that offer little resistance to flowing water. The ideal streamlined shape tapers to a point in the rear and reaches its widest point 36 percent of its length from the front. Many fish come close to being perfectly streamlined, but the shape is surprisingly rare among aquatic insects. An exception is the *Baetis* mayfly. This small, abundant larva is streamlined, not flattened like many other mayflies, and can swim through the current in short, fast bursts.

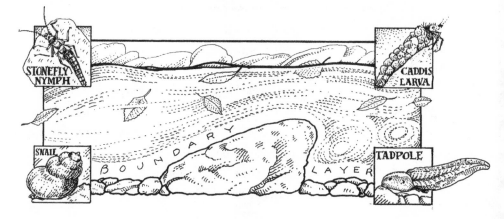

One of the most effective ways animals can stay put in fast current is with hydraulic suckers. The only invertebrates to use them are the larvae of Blepharoceridae, a family of flies found in most parts of the world. The six segments of their bodies are each equipped with a sucker with a soft outer rim and a dome-shaped piston in the center, which can be raised or lowered with muscle contractions. When the soft edges of the sucker come in contact with a rock or other solid material the piston is pushed down, forcing water out through a valvelike notch at the back of the sucker. When the piston is raised it creates a partial vacuum, holding the sucker firmly in place.

Various hooks, claws, or grapples are used by river-dwelling arthropods, members of an immense phylum that includes all the insects, arachnids, and crustaceans. Most common are small, well-developed hooks at the rear end of insect larvae, which they use to cling to rough surfaces. Other larvae have circlets containing hundreds of minute hooks on the ends of lobes extending from each side of their abdominal segments.

Many arthropods have solved the problem of staying put by secreting silk or sticky materials that anchor them to bottom. Some insects, like the torrent midge of Hawaii, spin tubes of silk around themselves, one end of which is firmly attached to a stone surface. The larvae of caddisflies spin a similar case when they pupate, but reinforce the relatively soft silk with bits of wood or tiny pebbles, constructing the equivalent of a brick house. These "cases" have the triple advantage of being sturdy, blending into the bottom, and providing ballast to keep the insect on bottom.

The larvae of *Simulium* are among the most successful of the torrent dwellers and have several adaptations for life in fast water. These flies, the best known of which are the biting black flies that are such a plague to mammals in northern regions, have large glands for producing silk. Black-fly larvae use this silk to build a supportive mat on submerged rocks, from which they hang by hooks arranged in a circlet at the rear of their bodies. A larva can inch across the bottom by bending its body in a U-shape, spinning silk from the glands near its head, releasing the hooks at the rear, and reattaching to the new pad of silk. If it should lose its grip and be swept away by the current it has a safety line: a single strand of silk attached to the rock that it uses to climb back to its original position.

Trout and salmon are splendidly adapted to life in fast water. Their streamlined bodies allow the current to flow easily over and around them and their powerful muscles and broad tails give them the ability to swim at high speed against strong current. Migrating salmon cannot afford to hold steady for long in rapids; they must push forward before they are too weakened to perform the rigors of spawning. In rapids it is easiest for fish to swim close to the bottom, dashing from one sheltered eddy to another, but if it is necessary to get over a waterfall the only way to do it is to jump. To ascend a falls a salmon leaps from the crest of a standing wave and tries to grab any advantage it can. By vibrating its tail rapidly as it launches into the air it is already swimming the moment it lands in the water near the crest of the falls. In this way it can surmount falls and other obstacles as high as 12 feet.

Other fishes tend to remain stationary near bottom or sheltered behind obstructions. Catfishes, sculpins, darters, gobies, and some suckers typically have flattened bodies and heads that taper back from the snout so the current pushes down and rushes over the back rather than lifting them from the bottom. Many have wide, outspread pectoral fins they use to anchor themselves. Others are equipped with such practical features as friction pads and ventral suckers. In the simplest adaptations, found in many darters and loaches, the front rays of the pectoral fins are thicker and stiffer than in other fishes and are placed well forward of the pelvic fins, so that all four fins act as friction plates on the bottom of the body. Many of the gobies, those widely distributed members of the family Gobiidae, are equipped with sucking discs where their pelvic fins have fused. One Australian species has such powerful suckers it can climb wet vertical faces beside waterfalls.

Some loaches, catfishes, and lampreys use their lips as suction discs to cling to rocks, a tactic that keeps them anchored in place but would seem to reduce their efficiency in feeding and breathing. Fish normally extract oxygen by allowing water to enter their mouths and pass over their gills, but those species that use their mouths to hold position on the bottom have lips that are divided into grooves or can otherwise allow water to enter. One fish in mountainous Thailand has adapted a double-channeled gill opening— one for allowing water in, the other for expelling it—freeing the mouth for the important work of clinging to the bottom and feeding. The *Garra* genus of Africa have thickened front lips and a buttonlike disc behind the mouth surrounding a hydraulic sucker. The tough front lip and sucker work together to allow the fish to crawl leechlike along bottom. The rock-climbing catfish of Colombia has a sucker mouth and barbed ventral fins it uses to shimmy up the perpendicular rock faces of waterfalls—climbing, according to one observer, a 20-foot falls in 30 minutes.

Rivers with gravel or cobblestone bottoms can usually support more life than those with bottoms of sand and silt. In addition to providing more surface area than a smooth bottom, stones create hiding places and—most important—a zone of still water along the bottom of the river. The cause of this zone of stillness, or *boundary layer*, is friction: Water flowing over a rough surface is slowed by the contact. When dyes are released into a stream and photographed, the boundary layer is clearly seen ranging from a few millimeters to a few inches above bottom. In this band of calm water it is possible for insect larvae, snails, fishes, and many other creatures to move with ease, while just above them every unanchored organism is swept away by the current. For the animals in this narrow world it must be like living on a planet where constant hurricane winds roar overhead.

10
OVER THE WATERFALL

Waterfalls are the flowers of geology. They are showy, extravagant, the liquid equivalent of morning glories in bloom and peacocks in full strut. They are majestic in the way that sunsets, mountains, thunderstorms, and redwoods are majestic, and majesty of that sort has a way of pulling us out of ourselves and humbling us and motivating us to travel hundreds of miles to stand as close to them as possible. We populate them with spirits, build observation decks overlooking them, and try—some of us, sometimes—to ride over them in barrels.

What is it about waterfalls that we find so irresistible? One possible explanation begins with the fact that when water is agitated as it rampages over waterfalls, crashes on seashores, or sprays through showerheads, great numbers of negatively charged ions detach from their host molecules and collect in the surrounding air. Research on the effects of these charged particles on humans has not been conclusive, but some researchers have suggested that negative ions reduce feelings of discomfort, anxiety, and aggression, perhaps by increasing the concentrations of neurotransmitters in our brains. If that's correct, waterfalls are an unlimited source of nonprescription Prozac, and we're drawn to them simply because they make us feel good.

NIAGARA FALLS
WINTER

THE CREEPING CREST of HORSESHOE FALLS

FALLS
ORIGIN—
7.5 MILES
DOWNSTREAM

1764

CIRCA 1890

1886

1995

TO LAKE ERIE

SPEED of CREST MIGRATION UPSTREAM ≈ 3'– 4½' PER YEAR

THE WALLS OF THE DOWNSTREAM GORGE REVEAL LAYERS
OF DOLOMITE AND SHALE THAT THE FALLS HAVE CUT
THROUGH ON ITS MORE THAN 12,000-YEAR-OLD JOURNEY

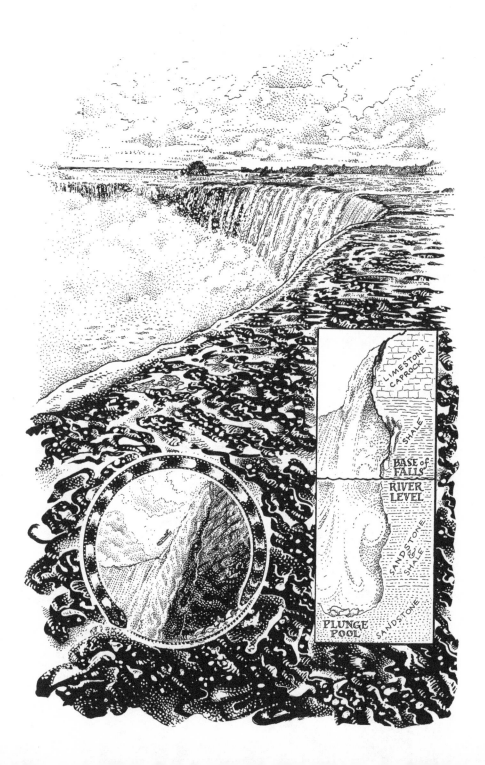

LIMESTONE CAPROCK

SHALE

BASE of FALLS

RIVER LEVEL

SANDSTONE & SHALE

PLUNGE POOL

SANDSTONE

Stand near the base of a falls, close enough to get drenched with spray and feel the ominous vibrations from tons of water bashing the earth, and you can become convinced that there is indeed something beneficial going on. It doesn't please everyone equally, but most people are mesmerized and delighted by the sight of water tumbling over a ledge, changing to foam and mist in the air, and crashing into a river.

Geologists, who by nature or training take a longer view of the world than most people, see waterfalls as brief and ephemeral features of the earth's surface. Compared to the slow lifting and eroding of mountains, waterfalls operate in fast-forward, their power focused and intense, their shapes changing much faster than most other features of the landscape. Even those of us with ordinary perspective can see that every waterfall is busily engineering its own destruction—retreating upstream and wearing steadily away until ultimately there will be nothing left of it.

Clues to the self-destructive history of waterfalls are often easily observed. Falls like Niagara and Yellowstone tumble into deep gorges because they flow over land composed of a cap of hard rock covering softer rock. As the hard cap wears away and breaks off, the falls migrate upriver. How fast they migrate depends on the composition of the rock, how much water flows, and the configuration of the crest of the falls. In the 12,500 years since the last glacier retreated, Niagara Falls have chipped their way upstream toward Lake Erie a total of about 7.5 miles, at a rate of about three feet each year. But that is an average rate. During some periods they have accelerated to four feet per year, while at other times the migration has slowed or stopped altogether.

Such variations occur in part because the composition of the rock varies—it takes longer to erode the cap where it is thick or made of harder rock—and because the amount of river flow fluctuates from year to year. At Niagara, where the falls are divided by Goat Island, most of the water goes over Horseshoe Falls, causing it to erode more quickly than American Falls. Horseshoe's rate of migration is affected also by the shape of the crest of the falls, which alternates between two configurations. When the crest is uniform and smooth the force of the water is distributed over a larger area and erosion slows. Periodically, however, a slab of the hard rock cap breaks off and the water at the crest is funneled into a V-shaped notch, increasing the force of the water at the center of the notch and speeding erosion. At the base of the falls plunge pools are formed by the falling water scouring the bottom of the river. As the falls move upstream the plunge pools are left behind like footprints to reveal where the falls lingered and where they hurried. Deeper, larger pools are left when the crest is smooth and the rate of migration slows or stops in what geologists sometimes refer to as "stillstands." Shallower plunge pools are left when the crest is notched and the falls have migrated more quickly.

Waterfalls are usually formed in one of three ways. First are those, like Niagara, where water passes over rocks of differing composition, with a layer of hard rock like dolomite overlying sedimentary rock that is more easily eroded. Second are those formed when plateaus are uplifted by tectonic forces, causing rivers to spill off them like water off a table. They include many of the oldest waterfalls, such as Angel Falls in Venezuela, the world's highest, and African falls like Kalambo (near Lake Tanganyika), Tugela (in South Africa), and Tisisat (at the headwaters of the Blue Nile on the Ethiopian Plateau), all of which were created during massive uplifts of land into plateaus and escarpments during the Tertiary Period, about 2.5 million to 65 million years ago. Related to escarpment falls are those that spill from "hanging valleys" that were eroded less quickly by glaciers than the deeper valleys they intersect, causing rivers to fall from the resulting precipices. They include Norwegian waterfalls that empty into glacial fjords, and some of the highest falls in Yosemite National Park.

The third category of waterfalls are those formed by "constructional processes." They can result when the flow of a river is blocked by deposits of calcium carbonate built up into what are sometimes called rimstone dams and falls. They can also be caused by lava, ash, and other debris from volcanoes; by ice dams, moraines, or ridges of sediment left behind by glaciers; or by debris from landslides and avalanches.

Differences in formation, terrain, and geology make waterfalls wonderfully varied. Some are cataracts, some cascades, some the treads and risers of watery stairways. In theory, water accelerates 32 feet per second as it descends over a falls; in the four seconds it takes to reach the bottom of a 250-foot falls it should be traveling at about 90 miles per hour. But in the real, imperfect world of waterfalls there are updrafts of wind that slow the water and ledges that reach out and divert it and variations in water flow that speed or slow descent. Some falls disappear into mist before they reach bottom or are dispersed so thoroughly by wind that almost no water reaches the ground below. Others freeze into icefalls in the winter and dry up altogether in the summer.

The most impressive waterfalls are those that combine great volume with great height. Any list of major falls has to include Victoria and Niagara, but the greatest of all in dimensions and quantity of water is Guaíra (also known as Sete Quedas), on the Paraná River at the border of Brazil and Paraguay. Its 18 cataracts combined are overflowed by an average of 470,000 cubic feet of water per second. Its vertical drop of 375 feet creates an earth-bashing spectacle that dwarfs Niagara's average flow of 212,200 cubic feet per second and maximum drop of 182 feet.

Niagara is nonetheless the champion falls of North America and one of the geological wonders of the world. The early Iroquois named it Nee-ah-gah-rah, "Thundering Waters," for the roar of the mighty spirit that dwelt beneath it, and sacrificed a maiden to it every year. The Senecas also believed Niagara was the domain of spirits, a benevolent one of thunder and lightning that lived in the mist at the bottom of the falls, and an evil one deep in the waters beneath that caused crops to fail and brought famine and disease. The evil spirit commanded an enormous serpent that lived in the Niagara River and Lake Erie and sometimes swam to the base of the falls, poisoning the water. When the good spirit of thunder and lightning killed the serpent with a lightning bolt, its thrashing death throes carved the horseshoe-shaped basin beneath the falls.

Rumors of Niagara's existence were detected long before the first Europeans traveled far enough into the American wilderness to hear its rumbling waters. The explorer Jacques Cartier heard rumors when he sailed up the St. Lawrence in 1534. A Jesuit priest, Father Raqueneau, wrote in 1648 that Indians told him there was "a cataract of fearful height" somewhere to the west, but he did not have the opportunity to trace the story to its source. A Franciscan missionary, Father Louis Hennepin, accompanied La Salle and his Indian guides to the falls in December 1678 and immediately set up an altar there and said mass. His later report, the first written description of Niagara, made the falls three times their actual height but did not exaggerate their power: "...more deafening than the

loudest Thunder; when one looks down into this most dreadful Gulph, one is seized with Horror."

British explorer David Livingstone likewise became aware of Africa's Victoria Falls before he saw them, but only because high plumes of mist were visible on the horizon 25 miles away while he was canoeing down the Zambezi River. The mist, as it turned out, rises about 1,000 feet above the falls and was the origin of the local name Mosi-oa-tunya, "The Smoke That Thunders." The mile-wide, 350-foot falls dumps as much as 270,000 cubic feet of water per second into a gorge of black basalt only 150 yards wide, forcing the river to turn at a right angle and rush to the end of the gorge, where it exits in a narrow gullet called "The Boiling Pot." Livingstone declared the falls "the most wonderful sight I had witnessed in Africa," and promptly renamed them in honor of the queen of England.

During high water, Aughrabies Falls on the Orange River, in Cape Province, Republic of South Africa, is more impressive even than Victoria Falls. Located in a remote searing land of salt pans and rocky desert, it plunges 480 feet into a granite chasm six miles long and up to 600 feet deep. At its base is a deep plunge pool, the reputed home of a fearsome mythological serpent known as "Groote Slang." The falls are a mere trickle in the summer dry season, but during the rainy season from October to March they produce a roar that can be heard 20 miles away, giving credit to their name, which is Hottentot for "The Place of Great Noise."

The world's highest waterfall is Venezuela's Angel Falls, which descends in a first stage 2,648 feet, strikes a ledge, then falls another 564 feet for a total of 3,212 feet. It is born on a high, uplifted plateau known by the local Camaracoto tribe as Auyán-tepuí, "Devil's Mountain," located not far from a similar table mountain described by Conan Doyle in his novel *The Lost World*. The falls' English name was given not because it seems to plummet straight from the realm of angels (though it does) but in honor of Jimmy Angel, a young American adventurer who was the first outsider to see the falls. He spotted it from the air in 1935 while flying over Venezuela's mountains in search of gold. He never found gold, but when he returned two years later with his wife they crash-landed on the swampy and boulder-strewn plateau near the brink of the falls and became famous for being perhaps the first people ever to look down those disorienting heights. The first to climb from the bottom to the top was Dr. Roger Latham, who led a scientific expedition there in 1968. Latham wrote:

> It's almost impossible to describe my sensation when I first walked out on the shelf overhanging Angel Falls... Suddenly there was the jungle more than 3,000 feet straight down...The awful height gave the ethereal sensation of standing suspended in space. It was like having the whole world suddenly opened up for inspection. But even more soul-stirring was Angel Falls itself. I lay down on the edge and inched my way out until my head and shoulders were over the edge... One lies there spellbound and awestruck at the overwhelming immensity of this panorama and perhaps for the first time realizes how truly big this world is.

The world's second-highest waterfall is Yosemite Falls, in Yosemite National Park California, which is composed of a freefall of 1,430 feet followed by a 675-foot cascade, then another freefall of 320 feet for a total descent of 2,425 feet. Also in Yosemite Valley is Ribbon Falls, which streams in a single 1,612-foot drop from a hanging valley. Both falls are best seen during snowmelt in May and June and often dry up completely in late summer. Another Yosemite falls, Bridalveil, is blown about so much by canyon winds that its water can hardly make it to the ground 620 feet below. It was originally known by two Yosemite Indian names: Wawona, "The Wind in the Water," and Posono, "The Spirit of Evil Winds."

Waterfalls have inspired much music, poetry, and art, and more than a little misrepresentation and hyperbole. A Scottish settler named Donald Sutherland announced late in the nineteenth century that while exploring New Zealand's South Island he had discovered the world's highest waterfall. His claim that the falls were more than 3,300 feet high earned him some fame—the falls are still known by his name—and a reputation as a liar: Later surveyors measured Sutherland Falls at a mere 1,904 feet. Franz

Schubert's "Gastein Symphony" was inspired by the 306-foot torrent at Badgastein Falls, Austria. The Norwegian composer Edvard Grieg found much inspiration in Vøringfoss, a 597-foot falls that brawls into a gorge above a branch of the Hardanger Fjord in Norway. Wordsworth's "sky-born waterfall" was Staubbach Falls, near Interlaken, Switzerland, which plunges more than 1,000 feet into an alpine valley.

Of course not everyone is inspired by waterfalls. Gertrude Stein said that Niagara was beautiful for about 30 seconds, though she may have been reacting more to the commercial enterprise surrounding the falls than to the falls themselves. Oscar Wilde said, "I was disappointed in Niagara—most people must be disappointed in Niagara. Every American bride is taken there, and the sight of the stupendous waterfall must be one of the earliest, if not the keenest, disappointments in American married life."

After English journalist George Augustus Sala visited Niagara in 1863, he wrote:

> These then were the famous Falls I had come so far to see... Well,
> I confess that as I stood staring, there came over me a sensation of
> bitter disappointment. And was this all?... There was a great deal of
> water, a great deal of foam, a great deal of spray, and a thundering
> noise. This was all, abating the snow where I stood and the black river
> beneath. There were the Falls of Niagara. They looked comparatively
> small, and the water looked dingy. Where was the grand effect—the
> light and shade? There was, it is true, a considerable amount of
> effervescence; but the foaminess of the Falls, together with the tinge
> of tawny yellow in the troubled waters, only reminded me of so much
> unattainable soda and sherry, and made me feel thirstier than ever.

Perhaps the stupendous scale of falls like Niagara overwhelm some people. Maybe expectations get overblown. Or maybe not everyone is affected in positive ways by negative ions. After all, researchers studying the effects of air ions on human behavior found that only about 30 percent of their test subjects were sensitive to them.

Still, I doubt if those of us who appreciate falling water are in the minority. Most people seem to react with wonder, delight, and astonishment when they see a great falls, even an exploited one like Niagara, and are likely to say something along the lines of what novelist Zora Neale Hurston said when she saw Niagara for the first time: "It's like watching the Atlantic Ocean jump off of Pike's Peak." Sometimes it takes a bit of hyperbole to do a place justice.

11

THE COLOR OF WATER

Lake Michigan is blue today, the color of robin's eggs and summer sky. Other days it has been darker or lighter, brighter or duller, showing blue in tints as various as a painter's palette. I have never seen it match the electric indigo of the Gulf Stream or the surreal festival blues of Caribbean flats, but at various times of the day and year I have stood on Lake Michigan's shore or flown over it or cruised it in boats and seen it as powder blue, baby blue, or the blue of jeans, both faded and new. Some days it is the quiet blue of herons, other days the vibrant blue of bluebirds and buntings. It can be periwinkle, phlox, or forget-me-not blue. It can be blueberry blue. It can be the hot blue at the base of a candle flame or the cool blue of menthol throat lozenges. It can be the color of bruises, of blue veins through milky skin, of blued hair, of shivering lips.

Blueness in water is usually but not always a testament to its cleanliness. Though the Dead Sea is vile-tasting and so salty that a swimmer emerges from it wearing a crackling white coat of minerals, it is colored a pleasing blue, as is Central Asia's largest salt lake, Koko Nor, the "Sky-Blue Sea." Dirty water tends to be green or yellow or brown, though sometimes water that appears dirty is perfectly safe to drink. Even the most pristine water will dull to gray when the sky darkens with clouds. During storms it will match the menacing black of thunderheads and become topped with furious banners of white.

The color of a body of water can reveal it to be shallow or deep, dirty or clean, temperate, tropical, or polar. An observant eye can watch a lake or ocean or river change in a few days from the crisp blue of winter skies to the smoky blue of summer haze to the laundered azure that follows a night of rain. As the sun swings from low to high to low again a body of water can reveal a dozen blues in a single day, a hundred blues during a year. Where the bottom drops abruptly from shallow to deep the delineation is marked with faded blue laid tight against deep cobalt, as if the colors were painted on with wide brushes. Water has a bit of chameleon in it. It can mimic the bottom—green of moss, yellow of sand, black of silt—or mirror the sky or steal the hues of trees and grasses along the shore. It might be green or yellow with algae, red with suspended clay, or white with finely ground stone. The geography of the bottom, the chemistry of the water, and the illusionist power of reflection are all revealed in color.

Some of the great variety of colors in water is due to the fact that every body of water is tinted with two kinds of color. *True color* is the "real" color of water, usually a result of the materials dissolved or suspended in it. It explains the tea brown of the Rio Negro in the Amazon Basin and the swimming-pool blue of Chile's Rio Puelo after it flows through copper deposits in the Andes. Rivers and lakes the color of café au lait are obscured—made *turbid* in hydrological terms—with suspended colloids so small that a single particle can take 50 years to settle through a foot of water. The particles of clay that give the Ohio River its earthy color have a diameter of only 1/1000 of a millimeter and would sink only about a foot per year if they were isolated in still water. Turbidity like the Ohio's can have natural or artificial sources. The beautiful Ohio praised in songs and poems two centuries ago was a very different river from the channelized and well-dredged highway for barges that now flows from Pennsylvania through Illinois. Now it is so dark with suspended clay and mud that light cannot penetrate. I once stood on a levee in Louisville watching a large tree tumbling with the current a stone's throwaway.

A Hydra head of black snaky roots lifted above the water, then sank. In a moment a length of black trunk appeared, rolling slowly like a capsized boat, followed by dripping branches and the reaching arms of secondary trunks broken off as if the entire top of the tree had been shattered in an enormous shredding machine. When each portion of the tree sank beneath the surface it disappeared entirely. There was no seeing through that water.

Lake Michigan, though it appears in many shades of blue, is not blue at all. The shades of blue we see are only *apparent color*, loaned to it by reflections on the surface or from the separation of light in its depths. By the glassful, Lake Michigan is clear.

When we draw clear water from a tap, scoop it from a mountain spring, or watch it spill in a crystal stream from a hose, it has neither true nor apparent color. It has no true color because suspended matter has been filtered out, has settled, or was never introduced. It has no apparent color because in small quantities it can't gather enough light to deceive our eyes. Take that same pellucid water and use it to fill an ocean or a lake and it becomes instantly colored. On an overcast day the water will reflect the white or gray of the sky. On a sunny day it will be blue for the same reason the sky appears blue. When sunlight enters water, its energy is quickly—and unequally—absorbed, or *quenched*. The long wavelengths at the red end of the color spectrum are absorbed first, always within the upper 30 feet. The blues and greens at the short-wavelength end of the spectrum are transmitted to the greatest depths.

How deeply colors penetrate depends on the clarity of the water and the angle of the sun. If there are many particles of sediment or microorganisms in the water, the "transmissivity" of the water is reduced and light cannot penetrate far. In clear water with high transmissivity, all colors are absorbed except blue, which is scattered and reflected throughout the water. The clearer the water, the deeper light penetrates, and the more pure its apparent blue will be.

Scattered light does not make water blue unless there is a lot of it. A glass of water remains clear because it does not contain enough quantity to scatter blue and absorb other colors. Light simply passes through. A swimming pool doesn't contain enough water either; if it appears blue it is because the bottom is painted or because the water has been treated with chemicals containing dyes the approximate color of pure water. It takes a good-sized river or lake to make water blue.

Reflections on the surface are the other source of apparent color. When the sun is directly above a body of water, about 2 percent of its light is reflected off the surface and the rest penetrates the water. When the sun drops lower in the sky, to an angle of 60 degrees, reflection increases to 6 percent. At 90 degrees—during sunrise or sunset—nearly 100 percent of the light is reflected. When light reflects off the surface of water it mirrors the colors of the world around it: the orange of sunset, the blue or gray of sky, the greens and yellows of forests on hills above the shore.

The transparency of water can be measured with a device called a *Secchi disk* (pronounced "seh-kee"), named for an Italian astronomer and Jesuit priest who invented it in 1865. It is a weighted white disk 20 centimeters in diameter, which can be lowered into the water on a line marked with measurements. The depth at which the disk disappears from view determines the "Secchi disk transparency" of the water, which is approximately the point at which only 5 percent of sunlight penetrates. Although there is room for error in the system (waves can make the disk hard to see, and it shows up better against a dark bottom than a light bottom), it is often used as a general measure of transparency. Lakes that are very turbid, such as Lake Texoma on the border of Oklahoma and Texas, might have a Secchi disk transparency of only a few inches. Muddy farm ponds or small fertile lakes with heavy blooms of plankton are likely to have a transparency of a few feet. In very clear lakes a Secchi disk can be seen 50 or 60 feet down. One of the clearest lakes in the world, Oregon's Crater Lake, has a disk transparency of about 130 feet.

COLORFUL NAMES

In Michigan there are five rivers named the Black. One of them is relatively clear and probably earned its name from the dark bedrock it flows over. The others are colored by suspended materials in the water. On overcast days or in the shadows of trees along the banks the rivers appear inky black, but in sunlight they are peaty brown, tinted with dissolved organic acids from vegetation decaying in bogs and swamps at the headwaters. Hold a glassful of their water up to a light and it is the color of weak tea.

A survey of any atlas reveals that hundreds of creeks, rivers, lakes, and seas throughout the world are named for colors—some for their true colors, some for their apparent colors, some for other colors altogether. Consider the White Sea off Russia's northwest coast. This nearly landlocked extension of the Barents Sea is entirely iced over from October to June and covered with rafts of drifting ice the rest of the year. Its name refers to the ice and snow on its surface, not its water, which tends to be a rich soupy olive.

The Black Sea was named centuries ago by Turks who took control of the region along its southern shore and declared it *Kara Dengiz*, the Black Sea, because it was a dangerous and inhospitable place swept by deadly dust storms. The water itself was once clean and blue enough to make it a popular resort destination of wealthy Russians and Ukrainians, but lately has suffered from eutrophication, the result of massive pollution with industrial and human wastes and farm fertilizers, and is now green with algae.

The Red Sea earned its colorful name because of its true—but only occasional—color. This long and narrow sea, stretching 1,300 miles from the Suez Canal to the Gulf of Aden and the Indian Ocean, is usually bluish green. Periodically, however, dense blooms of cyanobacteria containing red pigments fill the top layers of water. The same organisms bloom in the Gulf of California, which has sometimes been called the Vermilion Sea.

The Yellow Sea, a 600-by-400-mile inlet of the western Pacific located between China and the Korean Peninsula, is colored a vivid yellow, especially along the coast of China. It is there that rivers discharge silt washed down from China's interior. Of the rivers that empty into the Yellow Sea, the Huang or Yellow River is the most conspicuous painter of the sea. The Yellow River carries a heavy load of an unusual soil known as *loess*, a word derived from the German "to loosen," and first applied to a similar soil in the Rhine Valley. China's loess is a soft, fine sediment consisting of minute particles of quartz, mica, feldspar, and calcite mixed with clay dust. Blown from deserts in Central Asia, it was laid down in northwest China starting about three million years ago and now forms the deepest blanket of soil in the world, a fertile and easily cultivated mantle hundreds of feet thick covering thousands of square miles. Loess is accommodating stuff. When it dries it crumbles into talcumlike powder that is easily lifted by any persistent wind. When it is untended and unplanted it mixes readily with rain to form a slurry of yellow mud that flows obediently into any available watercourse. Most of those watercourses eventually find their way into the Yellow River.

In its headwaters the Yellow is a turbulent river, gnawing through thick beds of loess as it flows across the northern plateau. By the time it slows to cross the lower plains to the east it has become the muddiest of the world's major rivers, delivering about 1.5 billion tons of loess to the Yellow Sea each year. The mass of water flowing between its banks contains an average of 34 percent sediment. It is thick, "a rippling tide of liquefied mud, resembling thick lentil soup," according to one observer. Where the sediment has settled out in the lower river it has raised the riverbed more than 30 feet above the surrounding plains and has formed long natural levees along the banks. When unusually heavy rains fall in the watershed drained by the Yellow and its tributaries, the lower river surges, breaking through the natural levees and all the dikes and embankments constructed by humans to control the river. The resulting floods inundate land for hundreds of miles around and have been responsible for hundreds of thousands of deaths. It is why the Yellow has for centuries been called "The Sorrow of China."

The Hvítá (White) River in Iceland flows 80 miles from a glacier at the center of the island to the Atlantic on the southern coast. Like the Yellow River, it is named for its true color: a milky white from suspended glacial debris, or "rock flour," produced by the scouring of a glacier against rock. The rivers of Iceland fall into two distinct types. Some are spring fed, as clear as the nation's celebrated vodka, and are deservedly famous for their salmon and trout fisheries. Those that flow as meltwater from glaciers, however, can contain so much rock flour that they are nearly lifeless.

In the American west, the Colorado River deserves the "red" of its Spanish name. Or at least deserved it in the days before most of its water was diverted for irrigation and drinking water. The original, virgin Colorado carried astounding loads of sediment—it was the equal of the much larger Mississippi—colored predominately by the reddish sandstone dredged from the Grand Canyon.

Residents of the Amazon watershed in South America classify the tributaries of that greatest of rivers according to their color. Some are called *ríos blancos*, "white rivers," because they are colored a murky tan or yellow with suspended silt. Others are *ríos negros*, "black rivers," because they are tinted various dark shades by dissolved organic matter. The largest of the dark rivers, and perhaps the darkest, is Rio Negro, which enters the Amazon and for many miles downstream turns it into a divided river, black on one side, *blanco* on the other.

The Nupe people of Nigeria have a legend about the origin of the Niger River. Long ago, they say, their king was warned by an oracle that enemies would soon invade their land. The king gave a piece of black cloth to his daughter, who tore it in half and dropped the pieces to the ground, where they turned into a black river deep enough to protect the kingdom from invasion. The more prosaic explanation for the color of the Niger is that, like the Río Negro in Brazil and the Black Rivers in Michigan, it is stained with organic materials leached from decaying vegetation.

In the oceans, the most celebrated colors are the festive postcard blues of the Caribbean. Clarity and high salt content give other waters an even deeper, richer color. Fishermen have known for centuries that the more blue the water the more barren it is and have usually avoided such brilliant but relatively unproductive waters as the Gulf Stream, the Sargasso Sea, and parts of the Mediterranean. Colors other than blue often mean life—or death. The rich waters of the polar seas are olive green, the color of diatoms blooming by the billions. Elsewhere blooms of diatoms and other single-cell phytoplankton color miles of water with red, brown, yellow, or green, the colors coming from tiny grains of pigment within each cell which collectively stipple the water with their dominant color. When upwelling currents cause a bloom of certain species of dinoflagellates, they can produce the discoloration of water known as a "red tide." Sometimes the tiny dinoflagellates produce toxins that when ingested by shellfish can cause poisoning in humans. A few species also kill fish, shrimp, and crabs, or, indirectly, bottle-nosed dolphins when they eat menhaden containing heavy concentrations of the toxins in their livers. When the bloom of dinoflagellates ends they die and decompose in such numbers that much of the dissolved oxygen in the water is used up, killing bottom-dwelling fish.

DINOFLAGELLATE

CAUSES "RED TIDE"

Gonyaulex tamarensis

*

Late one summer day, as we walked the beach along Lake Michigan, my wife commented on the colors of the waves. I looked and saw gold and orange reflections of the setting sun, but nothing more. Gail suggested that I look deeper, that I was not seeing the subtleties of shape and color an artist sees. I concentrated and for the first time saw water as a painter must, as a problem of technique. The small swells were not objects with dimension, mass, and weight, but a complex arrangement of light and color, with blue, rose, and gold bars shifting back and forth across a flat background. How could such complexity be captured with a brush? John Ruskin complained that "To paint water in all its perfection is as impossible as to paint the soul." I understood why he thought so.

We watched the water until our eyes swam with waves, then played a game of listing as many blues as we could see. Gail saw phlox and iris, turquoise, sapphire, and lapis lazuli. I saw enameled cookware and razor blades and our oldest son's eyes. When the wind fell and the water calmed we collected a pile of the flattest stones we could find and threw skippers until the sun was gone.

12
ICE

On cold winter mornings before the ice comes, East Grand Traverse Bay is covered with wispy and shifting fingers of mist, like the luminous gases that dance above the surface of the sun. It's called sea smoke in the North Atlantic, steam fog in the meteorological texts. To most of us it is simply mist, caused when water evaporating from the lake condenses as it comes in contact with the frigid air above. Those restless clouds of droplets mean heat is being sucked from the lake.

This morning the sun shines orange through the mist but offers little heat. Already the bay's edges are fringed with ice so thin it looks like plastic wrap stretched over the shallows. From a distance it appears only as a slight altering of light on the surface of the water, a glistening reflection. Closer it is revealed to be a pane of ice as fragile as the finest crystal and decorated with the same delicate plumes and blossoms that scroll across frost-covered windows. When the water recedes slightly, tipped gently from one side of the bay to the other, the fringe of ice remains anchored in sand and pebbles, suspended an inch above the water like tiny glass decks for lakeside homes. Most days waves and sun take the ice away by midmorning, but every night more forms as a summer's worth of stored heat is released and the

PANCAKES

temperature of the water falls a little. If the temperature falls low enough, for long enough, ice will cover the entire bay.

East Bay is 20 miles long and 3 to 5 miles wide, with depths of nearly 600 feet. It takes a long time for so much water to give up its heat. The bay rarely freezes over before February, and many years it does not freeze at all.

But when the winter is a cold one the fringe ice comes and goes for weeks and the surface water seems to become thick and sluggish. It is not an illusion: As the temperature of water decreases, viscosity increases, making the water denser. Loose ice crystals form on the surface, creating faintly audible pinging and shushing as they collide in the swells along shore.

A lake freezes in either of two ways. When the air is very cold and still, with no wind to ruffle the surface, the entire lake can freeze over in a few hours. This typically happens at night, when temperatures are lowest and wind is less likely to interfere. If there is wind and waves the process is very different. The water wants to freeze, it *has* to freeze, but all it can make are isolated fragments of ice that bump against one another and become rounded by waves. As water washes back and forth across these pieces of ice, it smoothes them into roughly circular forms with raised edges, and causes them to grow until they are one to three feet in diameter. When the wind subsides, the separate floes of this *pancake ice* cement together into a solid but uneven blanket of soft ice, colored white with trapped air.

If the wind does not subside or if a storm blows large waves for days at a time, drift ice will surge and tumble along the shore and form spheres known as *ball ice*. The balls, some of them measuring three or four feet in diameter, roll back and forth across the bottom picking up so much sand and gravel that they sometimes sink.

Because the bay is large, it is rarely calm. Waves and wind collect the loose crystals of ice, form pancake ice, and, if the wind is steady, herd it in gently chiming swells against the windward shore. The southern end of the bay becomes covered with a frozen aggregate of pancakes pressed tightly together by the wind and welded by surface ice. It makes a rough, difficult surface, dense and uneven, with treacherous patches of dark open water that look like missing pieces from a jigsaw puzzle. Beyond the interlocked ice is water veiled with sea smoke.

Farther north on the bay, where open water comes in contact with the beach, waves shove drift ice to shore and coat it with spray, building an irregular thick fringe of ice six or eight feet high. This *foot ice* (or icefoot) lasts all winter, serving as a buffer between the waves and beach, growing during each storm and shrinking on warmer days. Because the water is a few degrees warmer, it erodes the front of the ice where it strikes, undercutting at the waterline and creating a balcony effect. Waves slamming against the ice often erode chambers deep within it. When waves strike the face of the ice, water shoots through the chamber with great force and finds a way to the surface, exploding in a plume of spray through a blowhole. The spray gradually builds a cone shaped like a volcano.

In liquid water, molecules are in constant motion, drawn toward one another by the weak covalent bonds of their positive and negative ends, colliding and ricocheting in constant, frenetic slam dances. They touch one another just long enough to keep the water together as a body, though not long enough to prevent random molecules at the surface from bounding away into the air as evaporation. When water is the temperature of the human body, only about 15 percent of the molecules are connected at anyone moment. These dalliances are unimaginably brief, about $1/100,000,000,000$ of a second each. But as the temperature of the water drops, the molecules slow their activity and draw together, reaching their closest proximity—and making the water densest—at about 39 degrees Fahrenheit. As the temperature drops even further the molecules are no longer able to break the hydrogen bonds connecting them to their neighbors, and the water begins to crystallize. At freezing the bonds weaken, causing the molecules to separate slightly, each of them locking at arm's length with four other molecules, all connected in a crystal lattice shot through with open channels. The spreading of the molecules as they freeze makes ice one of the few substance that are less dense frozen than liquid, and one of the few to expand rather than contract at freezing. The difference in density is only about 10 percent, but it is enough to make ice float—and that makes all the difference. If ice behaved like all other compounds and became more dense as it froze, its molecules nesting together in the tightest possible arrangement, ice would sink. In cold regions oceans and lakes and every living thing in them would freeze from the bottom up.

*

Late one night in February, after the wind dies and the water calms and the temperature of the air falls to single digits, ice crystals form in the water and connect in a floating latticework across the entire surface of the bay. It happens quickly. At dusk I stood on the shore and watched open water; when I return the next morning at dawn the water is coated with a thin covering of ice, the surface motionless and sheening with the reflected light of sunrise. All day the temperature is low and the wind light and the ice thickens, growing downward where liquid water comes in contact with solid water. By the second morning it is clear and hard, strong enough to hold my weight. A few ice fishermen make their way gingerly out to the depths where perch and whitefish are found.

At this stage the ice is sometimes called "black ice" because it is so clear that the dark depths of the lake show through. In the shallows the ice is as transparent as plate glass. On the bottom every pebble, drowned leaf, and zebra mussel is motionless and slightly magnified, as if frozen in amber.

Some years there is no clear ice. If the lake freezes during a heavy snowfall, the snow mixes with the new ice at the surface and forms a layer of inferior "white ice," which is weak and crumbly, shot through with tiny pockets of air. White ice is granular and opaque, while black ice is crystalline and transparent. Four or five inches of white ice will support an adult's weight, but you can drive an ice spud through it with a single stroke. Even when black ice has formed, snow can spoil it. The weight of a heavy snowfall presses down on the ice, creating pressure on the water below and forcing it to escape to the surface through cracks, where it mixes with the snow and forms slush. If snow continues, the slush beneath it freezes and creates more inferior ice. By the end of winter, if much snow has fallen, lakes tend to be covered with two or three feet of ice the color and quality of freezer crud.

Once ice covers a lake, some of the heat still stored in the liquid water is conducted away by contact with the ice and freezes, causing the covering of ice to grow from the bottom down. But as it gets thicker, and especially as it becomes covered with a layer of snow, it insulates the lake and prevents it from losing additional heat. Without such insulation, it would be possible for ice to form all the way to the bottom.

WHY THE ICE BOOMS

When a frozen lake makes new ice you can hear the growing pains. One night after East Bay has been frozen less than a week I walk down to the shore to watch the sky. A crescent moon and Venus hang low in the west, and above them stars pulsate through uneven layers of atmosphere. Clusters of lights in the villages up and down the far shore glitter like distant galaxies. About nine o'clock, after the temperature drops below 10 degrees, the ice begins to boom.

Freezing water expands irresistibly. The expansion is infinitesimal at the molecular level, but in the aggregate it combines to make a powerful force. It can shatter cast-iron water mains as if they were made of porcelain, heave saturated ground high enough to lift buildings off their foundations, and buckle asphalt highways. Entire mountains are broken apart and whittled down by the wedging force created when water infiltrates, freezes, and expands.

Send that expanding force across several miles of ice and the results can be explosive. Tonight the ice bucks and splits along the entire length of the bay, the noise it makes amplified by the drumhead covering the lake. The sounds vary. Sometimes they are continuous and steady, like an enormous bowling ball rolling across 20 miles of wooden flooring. Sometimes they are sudden and short, a series of abrupt electric cracks. There are random shots as piercing as artillery shells, followed by intermittent sonorous thumps, like giants stomping across the bay. Sometimes it is dull and thudding, sometimes metallic and reverberating. It can sound like distant jets or like noisily burning fires. One particularly loud peal rumbles down the bay and reverberates back and forth for a full 15 seconds. It ends with a sudden echoing belch, as if an enormous bubble rose to the surface and blew apart beneath the ice.

Most people, when they describe the noise of booming ice, say it sounds like thunder, which is fitting since the cracks that produce the sound are often as jagged as lightning bolts. A few years ago I stood on a frozen lake at dusk, when the air temperature was falling and the ice was beginning to grumble. There was a sudden sharp retort and the ice cracked in an erratic black line that traveled half a mile in a heartbeat and passed between my feet. I saw it coming: a zigzagging bolt of horizontal lightning that passed so quickly beneath me I could not react. If I had been sitting on a horse it would have been shot out from under me. I looked down and the ice on each side of the crack rocked lightly with my weight. A thin bulge of water seeped through the fissure, spread out, and froze.

*

Ice and snow are unstable. The bond between molecules in frozen water weakens when they stretch apart, just as the attraction between two magnets weakens as they're separated. Such weak bonds are sensitive to temperature, causing ice and snow to change as temperature changes. Ice locked deep inside a half-mile-thick glacier can remain relatively unaltered for thousands of years, but within the foot or two of ice covering a lake

CANDLES

it takes only a few days of warm temperature and bright sunshine to stir molecular change.

In March, after a week of mild temperature, the wind comes up and the ice on East Bay begins to move. It shifts, fractures, and sets floes the size of townships into motion. Some years storms shove the mass of ice to shore with enough force to be destructive. It digs furrows in the beach, uproots trees, and tears down jetties, breakwalls, and, on occasion, houses. This ice shove, or "ramparting," sculpts beachfronts, builds dunes, blocks stream mouths, and dredges rocks from the bottom and carries them far above the reach of waves.

This year the wind is insistent but not powerful enough to cause damage, and the floes of ice merely strain against each other. Already the warm days have had another effect: The covering of ice that stretched white and uniform has taken on the dark and mottled look of storm clouds. A quarter-mile out the ice is so dark it is probably not safe to walk on.

Near shore I cut a hole and lift out a block of ice the size of a birthday cake. It is hard and clear, vaguely blue, but the disintegrating crystal lattices have left it honeycombed with columns of vertical six-sided crystals, like lines drawn by a jeweler intent on cutting facets. I leave it on the ice in the sunlight and in 20 minutes the lines have become more distinct and the hexagonal crystals are beginning to slip apart. I kick the block and it splinters into hundreds of candles, each the length and shape of a transparent pencil. Around me, puddles on the ice are percolating downward in an audible trickle. It is clearly time to get off the ice.

Up the bay, as far as I can see, ice is grinding at pressure cracks, pushing upward against itself the way opposing continental plates push upward to form mountains. The ice groans and rumbles as it moves, rising gradually into ridges the size of motels, then into peaks that break beneath their own weight and slide down the slopes like big white chunks of broken highway. It is only March but there is a promise of May in the wind. I turn my face to the sun and try to imagine summer waves breaking gently on this beach. Gulls hunker on the ice by the pressure crack, their backs to the wind, waiting for open water.

OTHER FORMS OF ICE

Frozen water can take many forms. It precipitates as snow, sleet, or hail. It appears on the ground as *rime frost*, which forms when droplets that are supercooled—colder than 32 degrees Fahrenheit but not yet frozen—come in contact with an object, freezing and building into spikes and feathers of ice that grow facing into the wind. *Hoarfrost*, the frost that covers the windshield of a car on a cold morning, sublimates onto an object directly from water vapor, much like dew.

Needle ice forms on the ground when there is a very high moisture content in the soil, and especially when there is no snow cover. As the ice

FRAZIL

freezes and thaws thin spikes of ice are pushed up from the ground by the growth of new ice below. As it rises it takes anything in its way with it, slicing plants off at the roots, heaving the soil, rolling rocks out of the way. In the far north and at high altitudes, needle ice occurs on the perpetually frozen ground called *permafrost*.

Verglas, or *water ice*, forms when water freezes as it flows over rock, soil, or other surfaces. It can be a mirror-smooth slick on the face of a mountain or the treacherous patch of shining ice on the surface of a highway.

Rivers can form several kinds of ice. The surface of a river does not freeze as quickly as the surface of a lake because current causes a constant mixing of water, making temperatures the same from bottom to top. That mixing means that the water of the entire river must reach the freezing point, unlike a lake where only the surface layer needs to get that cold. Mountain streams occasionally freeze solid all the way from the surface to the bed, though water usually continues to flow between stones on the bed and in deep pools. When the surface of a river freezes over, the water level sometimes drops as upstream sources freeze, leaving the ice suspended as a bridge.

On cold, clear nights when the air temperature is below about five degrees Fahrenheit and the surface of the river remains unfrozen, ice can form underwater. Such subsurface ice results when fast current keeps a river ice-free even after the water temperature drops below freezing. Ice in such supercooled water can take two forms. *Frazil* or slush accumulates as ice crystals form or when large amounts of snowfall, transforming the river into a flowing daiquiri. During extremely cold temperatures, *anchor ice* can form on the bottoms of rivers, first in riffles on the upstream faces of stones, then across most of the bottom. It usually forms at night, when radiant heat escapes from rocks and other objects on the bottom. During the day the ice is often warmed enough by the sun to break free and tumble away downstream, causing surges of current. Anchor ice rarely occurs in deep water, but in polar regions it accumulates on ships' anchor chains (which are effective heat conductors), sometimes lifting anchors off the bottom and setting ships adrift.

SEA ICE

As freshwater cools it becomes denser until it reaches about 39 degrees Fahrenheit. If it continues to cool beyond that temperature, the molecular structure of the water becomes more crystalline and less tightly packed, making it less dense until it forms ice and floats to the surface.

But the freezing point of water varies, depending on such factors as atmospheric pressure and purity. Salt water behaves a little differently as it freezes. It continues to grow denser until it reaches the freezing point, which, because salt water contains so many dissolved salts and minerals, is as low as 27 degrees Fahrenheit. As the water crystallizes into ice, dissolved salt is forced out into the liquid water below, where because of its density it sinks and is replaced by less salty water that rises to the surface and freezes to the underside of the ice. If the surface ice has formed slowly, most of the dissolved salts drain away. If it freezes quickly during extremely cold

weather, brine is trapped in voids between the crystals of ice. As the ice ages the brine escapes until the ice is fresh enough to drink when melted.

When sea ice begins to freeze, ice crystals gather in a loose gruel of frazil. As frazil thickens, the water becomes souplike and the surface looks somewhat like congealing grease tinted the color of lead or gray steel (our words slush and sludge may be derived from the Russian *salo*, which means "grease"). This grease ice freezes into a one- or two-inch layer of elastic ice called *nilas*, which covers the rising and falling swells like a dark cloak. If broken up by wind and waves, nilas becomes the raw material for pancake ice. When unbroken nilas reaches a thickness of about four inches (strong enough to support a person's weight) or when pancake ice congeals and freezes solid, it is known as young ice. Areas of young ice are frequently saturated with seawater that freezes and helps cement the whole. The water for a foot or so beneath it is saturated with ice crystals that adhere to the ice and make it grow thicker from the bottom down. By the end of a winter the young ice—or first-year ice, as it is sometimes called after it turns opaque or white—has usually reached a depth no more than about 6.5 feet. The depth of the ice is limited because thick ice becomes an efficient insulator that prevents further heat from being extracted by conduction from the warmer water beneath. When no more heat can be extracted the water will not cool enough to freeze.

First-year ice that does not melt during the summer becomes hard, blue, second-year ice. If it lasts beyond the second winter it becomes polar pack ice and continues building in thickness until it reaches 10 to 16 feet deep.

Where sea ice accumulates on the surface along coastal regions, especially in bays and other protected places, it forms a strong, thick, immovable layer that gradually spreads outward and grows thicker as the winter progresses. The portion of this *shorefast ice* (or simply fast ice) that comes in contact with the shore and is not subject to the rising and falling of tide is called *ice foot*.

When wind or current causes fast ice to drift away from the shore it becomes *pack ice*. Early Arctic explorers defined an *ice field* as an area of pack extensive enough that its edges could not be seen from the masthead of a ship. Pack smaller than an ice field, down to about one-third of a nautical mile in diameter, is called an *ice floe*. Pieces smaller than a floe are sometimes referred to by the French word *glaçon* ("an individual piece of ice").

The surface of polar sea ice is rarely as uniform as the surface of freshwater lake ice. Wind and upwelling currents create areas of perennially open water known as *polynyas* and long channels of open water known as *leads*. Leads slam shut as wind and current drive ice fields against one another, crushing them together and forcing their edges to rise in ridges known as hummocks. When a great number of *hummocks* are pressed together into a free-floating mass they become a *floeberg*. When hummocks break up into glaçons they are sometimes called *growlers*.

The amount of sea ice covering the oceans of the world varies from season to season. It arrives every winter and breaks up every spring along the shores and in bays in Alaska, much of Canada, Russia, and Scandinavia, and parts of the northeastern United States, and remains as a permanent cover around Antarctica and in the central portions of the Arctic Ocean. At both poles, raft ice many years old, composed primarily of fields of hummocks, form very solid packs that grow to be enormous. The Arctic ice doubles in area every winter, from summer's 3 million square miles to 6 million square miles. In Antarctica, the pack ice surrounding the continent covers about 1 million square miles in summer, but expands every winter until it covers some 8 percent of the Southern Hemisphere—about 7.7 million square miles of ice, an area so large that it affects the climate of the entire world, reflecting sunlight back through the atmosphere, trapping warm water inside the oceans, and creating powerful currents as high-density brine sinks beneath the ice.

GLACIERS, ICE SHEETS, AND ICEBERGS

Ice is easy to make. If our planet happened to follow an orbit slightly farther from the sun most of our water would be permanently frozen. When climate swings to the cool end of the pendulum and the average temperature goes down a few degrees, ice spreads across the earth like a fungus. During the last glacial epoch, which ended about 12,000 years ago, enormous glaciers covered most of the landmasses.

Remnants of those glaciers still exist in the polar regions and in high altitudes—wherever snow and ice accumulate faster than they melt. Permanent snow and ice are found at elevations ranging from sea level at the poles to approximately 16,400 feet above sea level at the equator. The

polar regions, because they receive only indirect sunlight during the summer months and none at all in the winter, accumulate ice at a much faster rate than it can melt and as a result are capped with massive amounts of ice and snow. In fact, the ice sheets and glaciers that cover most of Antarctica and Greenland and form a permanent cover over ocean waters near the poles account for about 80 percent of all the freshwater on earth. If all the polar ice were to melt, the oceans would increase in volume by about 2 percent, raising their levels approximately 300 feet.

Ice sheets (also called ice caps) are glaciers that are not confined to valleys. They form as relatively small amounts of snowfall accumulate over long periods of time, the weight pressing the snow beneath and changing it gradually to ice. Ice sheets move, so water frozen there is not frozen forever. But the pace is slow: The residence time of water in glaciers is approximately 10,000 years.

Glaciers form when the amount of snow that falls is greater than the amount that melts. In places like Antarctica and Greenland, and in high passes in the Alps, Rockies, and other mountain ranges, this snowpack builds gradually, year after year, decade after decade, century after century. As it builds, the snow beneath becomes compacted by the weight of the snow above and changes to ice. In time, the ice can become thousands of feet thick.

Glaciers move because a thin layer of liquid water forms at the bottom of the glacier. This water, produced from latent heat in the ground that melts ice or as meltwater trickling down from the surface, serves as a lubricant that allows the glacier to creep downhill at rates varying from a few inches to many feet each day.

Icebergs are born when pieces of glaciers break off into oceans or seas. If this calving is from a high, relatively narrow glacier in a valley, as are common along the coasts of Alaska and Greenland, the resulting high icebergs with their elaborate towers are known as *castle bergs*. When continental ice sheets such as those that cover Antarctica break off they form huge, flat *tabular bergs*. Arctic icebergs are driven south by currents into the North Atlantic (it was such an iceberg that sank the Titanic), while Antarctic tabular bergs form massive islands of ice that drift, sometimes for years, in the waters near the polar continent.

In recent decades, as the global climate has warmed, polar ice has undergone dramatic changes. A 2012 study published in the journal *Science* reported that Greenland, West Antarctica, and the Antarctic Peninsula were losing hundreds of billions of tons of ice per year. Only East Antarctica was immune; it was gaining about 14 billion tons per year, probably because snowfall rates in that region were climbing as the climate changed. One result of polar melting is faster-than-anticipated increases in sea levels, to about 3.2 millimeters a year. Scientists project that if the trend continues, millions of people will be displaced from coastal regions, many species of animals and plants will be put at risk, and hurricanes and other storms will be "supercharged" by warming waters. Eventually it could be catastrophic: If just one-tenth of the ice in Antarctica were to melt it would cause oceans around the world to rise approximately 30 feet.

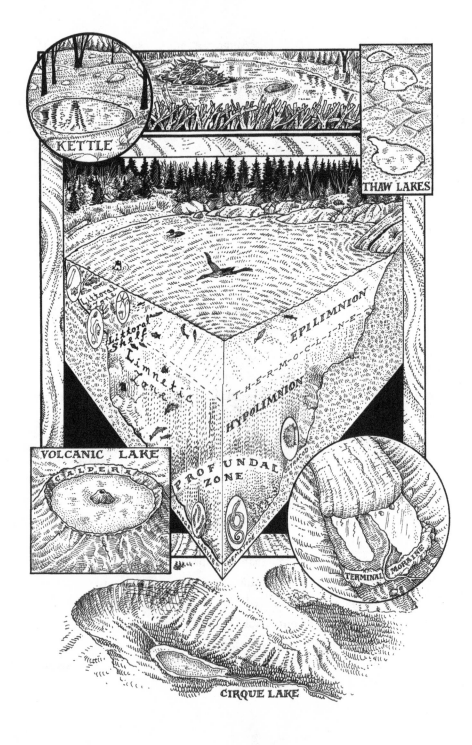

KETTLE

THAW LAKES

VOLCANIC LAKE

CALDERA

EPILIMNION

THERMOCLINE

HYPOLIMNION

Littoral

Littoral Shelf

Limnetic

Lotic

PROFUNDAL ZONE

TERMINAL MORAINE

CIRQUE LAKE

13

SKY WATER: PONDS AND LAKES

When clean water becomes as precious as gold, Lake Superior will be the mother lode. Its 31,700 square miles give it the greatest surface area of any freshwater lake in the world and the greatest volume of freshwater in the Western Hemisphere. Combined with the other four Great Lakes it forms the largest network of freshwater lakes on earth. It is so large it exceeds our perceptions. It is more than just a body of water, it is a presence, a shaper of geography and a creator of climate. For two years I lived near it and was constantly astonished by its power and the whimsy of its moods, glass-calm one day, deadly violent the next, always unpredictable. It's a formidable, beautiful, and unusual lake.

Or not so unusual. In the lexicon of limnology, the science of freshwaters, a lake is defined simply as a body of nonmarine water. It's a broad definition. It groups Superior and other freshwater seas with alpine ponds. It includes freshwater lakes and saline lakes. It includes lakes a few feet deep and lakes a thousand feet deep, those with water clear as springwater and those with water dark as coffee. Sometimes the definition is merely regional, as in the ponds of Maine, which can be ten miles long and hundreds of feet deep. "Sky water," said Thoreau, watching Buddha clouds on the surface of Walden, a pond which in most parts of the world would be a lake. In Newfoundland nearly every lake is called a pond and in Wisconsin any pond large enough to hide a bass or walleye is called a lake. But to a limnologist, they are all, large and small, lentic habitats—from the Latin *lentus*, "sluggish"—and subject to the same laws and tendencies.

Superior is like the majority of ponds and lakes in that it was formed by glaciers during the Pleistocene epoch, which lasted from about 400,000 years ago to about 10,000 years ago. The glaciers disrupted river systems, changing the courses of some, eradicating others. The southern boundaries of the ice became vast dams for north-flowing rivers, building reservoirs that overflowed and ran laterally along the edge of the ice, forming new river courses. In North America those well-worn courses are still followed by the Ohio and Missouri Rivers.

The glaciers made lakes by the thousands across Finland, Sweden, Norway, Scotland, England, Canada, Alaska, the northern tier of the United States, and in southern South America and New Zealand's South Island. The Great Lakes in North America and the Baltic Sea in northern Europe were formed by a combination of glacial scouring of existing valleys and the sheer weight of the glaciers. The ice was nearly two miles thick in some places, and so heavy that it depressed the ground beneath it as much as 2,000 feet. As the glaciers retreated, the ground behind rebounded in a slow lag effect that caused the depressions to slope toward the ice and fill with meltwater to become permanent lakes. Among the major lakes formed along the margins of retreating ice were Canada's Lake Athabasca, Great Bear Lake, and Great Slave Lake. Glaciers were also responsible for a number of other types of lakes. As they melted they left behind many long, narrow basins that had been bulldozed beneath them as they advanced. If the lower ends of those basins were blocked by terminal moraines—the piles of rock and soil shoved ahead of the glacier as it advanced—the moraines acted as dams to form elongated lakes. New York's Finger Lakes and some of the largest lakes in the European Alps, including Como, Constance, and Lucerne, were produced this way. Small lakes known as kettles formed when blocks of ice buried in the drift for tens or hundreds of years melted and the soil above them collapsed. Thousands of them are scattered across central Canada, Minnesota, Wisconsin, and Michigan.

Small, shallow ponds called *thaw lakes* are found in the permafrost-prone tundra of northern Europe and North America. They form when the surface of permanently frozen ground thaws each summer, eventually causing it to sag into shallow lake basins. Hundreds of them along the coastal plain of New Jersey probably formed when glaciers lurked just north of the region.

Glacial erosion in mountains sometimes forms amphitheater-shaped valleys called *cirques*. If debris blocks the valley's lower ends and they fill with water they become *cirque lakes*. These steep-sided lakes, each with a single outlet stream, are common in the northern Rocky Mountains, notably in Glacier National Park in Montana. Downstream from them are sometimes found series of smaller lakes in descending chains. Because they resemble strings of rosary beads they are known as *paternoster lakes*.

Pleistocene glaciers were indirectly responsible for many lakes that formed south of the actual ice. Increased rainfall and flooding rivers filled previously dry basins in arid regions, creating *pluvial lakes*, many of which dried up as temperatures rose and rainfall declined in the following millennia. Twenty thousand years ago Death Valley was a pluvial lake hundreds of feet deep. The only water remaining from another, Utah's Lake Bonneville, is in Great Salt Lake.

Glaciers are responsible for most of the lakes in the world, but certainly not all of them. Limestone caves collapse to form sinkholes that fill with groundwater, forming small lakes. Beavers dam creeks to make temporary reservoirs; if the dams are sound enough to last decades they become anchored with trees and create permanent lakes. People build reservoirs, farm ponds, and industrial lagoons. Wind and waves along ocean coasts form sand spits and barrier beaches that isolate estuaries and bays into lakes. Something similar happened at the *drowned rivermouths* along the east shore of Lake Michigan, where the prevailing west winds have pushed rivers back and formed natural lakes a short distance inland.

Lakes can also result from the emergence and subsidence of landmasses or by changes in sea level that leave a portion of a sea isolated and landlocked. Some evidence suggests that Florida's Lake Okeechobee, the largest United States lake south of the Great Lakes, is a ten-million-year-old remnant of an ancient sea floor.

In Quebec, Lake Chubb's shape betrays an extraterrestrial source: It is an eerily perfect circle, two miles across, formed millions of years ago by the impact of a meteorite. Another impact crater, Arizona's Meteor Crater, is dry now but was a pluvial lake 10,000 years ago.

Inactive volcanoes can become lakes if their calderas, or craters, fill with precipitation and meltwater. Crater Lake in Oregon filled 2,000 years ago following a long series of eruptions of Mount Mazama. At 1,932 feet deep, it is the second-deepest lake in North America (after glacially produced Great Slave Lake, which is 2,015 feet deep). Fed only by precipitation and melting snow, its water is incredibly clear, giving it a rich sapphire color. Lake Mashu, on the Japanese island of Hokkaido, is a smaller but equally lovely caldera lake with a maximum depth of 695 feet. The water in both crater lakes is so transparent that objects can be seen on bottom 135 feet below their surfaces.

Many large, deep lakes were formed by faults in the earth's crust. They include some of the oldest lakes, such as Lake Baikal in Siberia and Lake Tanganyika in East Africa, which formed 20 to 25 million years ago, but also include youngsters like Reelfoot Lake in northwest Tennessee, which appeared in a rift during the New Madrid earthquakes in 1811 to 1813.

Regardless of how they form, once basins become lakes—filled by water from precipitation, runoff, groundwater infiltration, or all three—they become lake or pond ecosystems. What kind of ecosystem and what kind of life they support depend on many factors, but all lakes that support life take part in the fundamental flow of energy from sun to plants to animals. The food chain (or web, if complex) begins with submerged aquatic plants and the tiny drifting algae known as phytoplankton, and proceeds to tiny plant-eating zooplankton, insects, minnows, large fish, and all the other producers and consumers living in the lake.

Lakes can be young, middle-aged, or old. Young or *oligotrophic* lakes have bottoms of sand or rock too clean of sediments for plants to take root. Without phosphorus and other nutrients to support plant growth their waters tend to be relatively free of algae and other organisms and therefore very clear. Though attractive to look at, such lakes are probably quite unproductive of life.

In time, however, enough organic material finds it way into a lake to build a layer of sediment and provide nutrients to support a healthy and diverse ecosystem. A lake is then middle-aged, or *mesotrophic*.

But too many nutrients lead to old age. A lake too enriched with mineral nutrients and organic material is said to be *eutrophic* and is at risk of dying. Unfortunately, eutrophication is not always a natural process. It can be speeded by the introduction of nitrates and phosphates from lawn and agricultural fertilizers, industrial waste, and sewage. Such was the case in Lake Erie in the 1950s and 1960s. Industrial pollution and runoff from farmlands and cities caused the rate of eutrophication to be speeded up by an estimated 50,000 years. With fertility accelerated, algae and other plants grew at a tremendous rate. For a time the bloom seemed an advantage. Organisms from the bottom to the top of the food chain became abundant, healthy, and well fed, and the fishing was better than anyone could remember. But when eutrophication continues plants begin dying in such numbers that their bacterial decomposers use up oxygen faster than it can be replaced by photosynthesis. Eventually Lake Erie became too oxygen-depleted for most organisms to survive and the entire ecosystem collapsed. Lake Erie is much cleaner now than it was in the 1960s and again supports a healthy fishery, but algae blooms in the summer of 2013 were reminiscent of the 1960s. Still, Erie is fortunate; many lakes that undergo accelerated eutrophication never recover.

Lakes have three ecological zones inhabited by different kinds of organisms. The *littoral zone* is the shallows, where enough light penetrates the water to allow rooted plants to thrive. It is the most densely inhabited and complex of the zones, home to numerous plants, plankton, snails, shrimps, crayfish, burrowing and swimming insects and their larvae, small fish, and large fish.

The *limnetic zone* is the zone of open water. It extends beyond the littoral zone, from the surface down to the point where light no longer penetrates. It is home to drifting and swimming organisms, from plankton to fish to piscivorous predators such as seals, otters, kingfishers, grebes, and loons.

The *profundal zone* is the lake bottom, where the only nutrients come from organic material raining down from above and settling on the bottom as sediment. It is too dark there for rooted plants, but the sediment is inhabited by bacteria, worms, nematode worms, ostracods, larvae of midge flies, and (in some waters) deep-dwelling mollusks, all of which feed on detritus, other insects, or deep-dwelling algae.

In temperate regions, large lakes always undergo thermal stratification. In spring and summer, warm temperatures and long days heat the surface water, making it less dense than the colder water below. This layering of water by temperature produces a warm surface layer called the *epilimnion* above a cold layer known as the *hypolimnion*. The boundary between the layers is the *thermocline*.

The epilimnion zone is a busy place. Surface water mixing with air and the photosynthesis of algae give it the highest concentrations of oxygen in the lake and allow it to support the greatest concentration of plants and animals. It is also where current and wind are active, keeping the water circulating.

The hypolimnion, in contrast, is cold, dark, and still. Its nutrients come only from organic material drifting down from the water above, and its oxygen is limited to what was already dissolved in the water when it stratified. By late summer, bacteria on bottom have used so much oxygen that the supply can be depleted.

If stratification is permanent in a lake there is no way to replenish the hypolimnion with oxygen and it will be nearly lifeless. But where there are hot and cold seasons, lakes undergo periodic *overturns* that mix the water, bringing dissolved nutrients to the top and carrying dissolved oxygen to the bottom. Overturn occurs in the autumn when the epilimnion cools to 39 degrees Fahrenheit, the temperature at which water is most dense, and the surface water sinks. Another overturn occurs in the spring when the cold surface water warms to 39 degrees. Again it sinks and again there is a mixing of water.

Anglers in Lake Michigan discovered years ago that in the summer another kind of thermal division occurs. Ten or fifteen miles from shore they found a point where they could pass from fairly warm surface water to very cold water in the space of a few feet. This *thermal bar* occurs in a distinct line running parallel to shore and is easily spotted because it often traps a "scum line" of drowned insects, dead baitfish, floating plants, and other debris, all of which attract trout and salmon. Thermal bars are found only on very large lakes, where shallow inshore waters warm much more quickly than the deep, cold offshore waters. A vertical temperature break forms, remaining in place until strong winds mix the water.

VERY LARGE, VERY OLD LAKES

The nature of a lake depends on many characteristics of geology and environment, but no factors make a bigger difference than great age and great size. One of the biggest and oldest of all lakes, Russia's Lake Baikal, is also one of the most unusual. Though its surface area of 12,000 square miles is only about 40 percent of Lake Superior's, it is the deepest of lakes, with a maximum depth of 5,314 feet (four times deeper than Superior). Such depth gives it the greatest volume of any lake, and about one fifth of the entire reserve of standing freshwater on earth. It is long and relatively narrow, about 30 miles wide and 395 miles long, bounded on all sides by steep mountains that plunge directly into the lake. As a result Baikal has few shallows.

Lake Baikal's isolation from other lakes and its 20-million-year age have allowed its flora and fauna to evolve into many species found nowhere else. Approximately 35 percent of Baikal's plants and 65 percent of its animals are endemic. The 975 species of endemic animals so far identified have evolved so far from their ancestors that they are placed in at least 87 genera and 11 families and sub-families of their own.

The most famous of the lake's unique animals is the Baikal seal, the world's only freshwater pinniped. It is a small seal, about four feet long and weighing 110 to 290 pounds, found mostly in the northern and central portions of the lake, especially on a cluster of four rocky islands known as Ushkany. Like many seals, it bears its young on the ice in February and March. Biologists believe it is descended from ancestors that migrated to the lake from the Arctic during glacial periods.

Baikal's most prolific endemic species are the freshwater shrimps. Of the 35 genera in the lake, 34 are found nowhere else, and the 255 species total about a third of all the freshwater shrimps on earth. Like the cichlids of Lake Victoria they are found in a variety of habitats throughout the lake, though many of them overlap.

The bulk of the endemic species in Baikal are found in very deep water, where living conditions have changed little for millions of years. In this freshwater abyss there is no sunlight and thus no photosynthesis, so the food web commences with detritus originating from the upper layers of the lake. On the bottom are found only two types of animals: those that feed on the rain of organic materials, and those that feed on other animals. In those depths are blind pinkish-white shrimp that navigate only by feeling their way with long antennae, and which are preyed on by two species of small, nearly transparent fish found only in Baikal. Another endemic fish, the omul, lives in shallow regions of the lake and is an important food source for seals. The salmonlike omul is also the species most sought by humans, making up nearly three-quarters of the commercial catch in the lake.

The world's second-deepest lake, at 4,822 feet, is Africa's Lake Tanganyika, which filled a 45-mile-wide and 410-mile-long trough in Africa's Rift Valley about 25 million years ago. It too has a remarkable endemic fauna, though unlike Baikal most of its unique species are found in the shallows and surface waters of the littoral and pelagic zones. Oxygen never penetrates below about 650 feet in Tanganyika because the water above is permanently stratified. Its profundal zone contains virtually no oxygen and is inhabited only by anaerobic bacteria.

Seventy-five percent of Tanganyika's fishes are endemic species. Among them are the herringlike Tanganyika sardines, abundant, four-inch dwellers of the pelagic zone and the lake's primary commercial fish. They are still fished today the way they were when David Livingstone visited the lake in the mid-nineteenth century and witnessed strange lights dotting the surface at night. Fires on platforms attached to fishermen's canoes attracted schools of sardines to the surface where they could be captured in nets. Kerosene lanterns have since replaced the open fires, but otherwise the same methods are used.

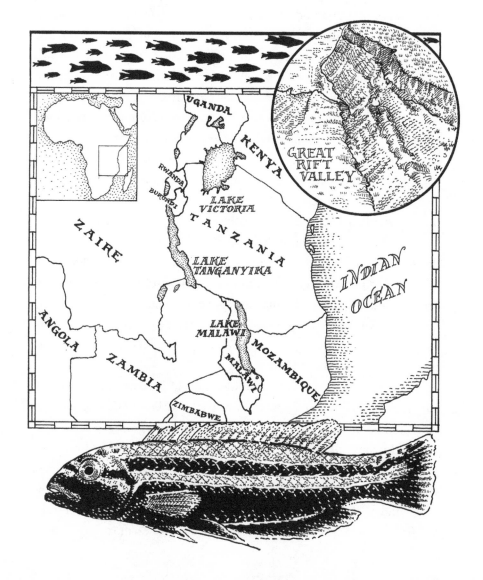

Other endemic species in Lake Tanganyika include two snakes, one a cobra that lives along shore and feeds at night on fish, the other an aquatic species that hunts sardines in the surface waters at mid-lake. Insects of note include a species of flightless caddisfly that skims the surface of the open lake like a giant whirligig beetle. Similar flightless caddisflies are found on Lake Titicaca in the Andes and Lake Baikal.

Tanganyika is a remarkably clear lake—so clear that early biologists assumed it must be unproductive. Clear water means few algae, and without that essential foundation to the food web it seemed unlikely that the lake could support much life. Yet the commercial fishery thrived. Tanganyika algae are unusual in that their population at any one time is relatively low yet their life cycle is extremely fast. They grow, reproduce, and die rapidly, making the transfer of energy from algae to zooplankton to fish extremely efficient, and producing an abundant food source in spite of a low standing crop. The sardines also find a reliable source of food when they migrate to deep water during daylight hours. At the thermocline, where oxygenated and deoxygenated waters meet, are large numbers of protozoans that feed on deep-dwelling bacteria.

Lake Malawi is the fourth-deepest lake in the world, with a maximum depth of 2,316 feet. It lies near the southern end of the Rift Valley, about 200 miles southeast of Lake Tanganyika, and is similarly shaped. It too is very old and supports many endemic species of fauna, especially fish. Also like Tanganyika, it has a productive surface zone and depths filled with oxygen-depleted water. Oddly enough, none of Lake Malawi's endemic fish species are found in Lake Tanganyika.

Of the 245 species of fish so far identified in Lake Malawi, about 200 are cichlids, 95 percent of them endemic. Cichlids from Malawi have long been popular among home aquarium enthusiasts for their colors and variety. In the wild they are remarkably specialized. Some feed on blue-green algae growing on the rocks, others on other algae, insect larvae, small crustaceans, or other fish. Some feed only on the tops of rocks, others in the cracks and crannies between them. A few species apparently feed on scales and fins they nibble off other cichlids, and a few prey on the young of other cichlids by sucking them from the mouths of their brooding mothers. Still other species live in the open waters of the lake and feed on zooplankton.

Another African lake with a large and diverse population of cichlids is Lake Victoria. Second only to Lake Superior in surface area, its 26,500 square miles make it the largest lake in Africa and the dominant feature of Africa's Rift Valley. Its basin formed millions of years ago when the eastern and western branches of the Rift Valley opened up, leaving a plateau between them. The center of the plateau sagged, causing rivers to reverse direction and empty into the basin. Though not as deep as Tanganyika or Malawi (about 270 feet), in the recent past it supported at least 400 species of cichlids, most of them endemics that probably evolved from a few species that lived in the lake's tributary rivers. The speed at which cichlids evolve into new species can be seen in a small lake adjacent to Victoria, Lake Nabugabo, which was separated from the main lake by a sandbar about 3,700 years ago. Of the nine cichlids found in the smaller lake, five are endemic to Lake Nabugabo, though closely related to species in Victoria. The cichlids of Victoria are currently threatened by an alien species, the Nile perch, which was introduced into the lake in the 1950s. The large and voracious perch has apparently wiped out about 200 species of cichlids in what one scientist has called "the greatest vertebrate mass extinction in recorded history." The lake has also become dangerously eutrophic due to runoffs of sewage, fertilizers, and industrial wastes, spurring international efforts to halt pollution and retain what is left of the lake's biodiversity.

The deepest lake in the British Isles is Loch Morar, at 1,017 feet, but Loch Ness has a greater average depth (433 feet) and more surface area. Both are located in long, narrow basins in northern Scotland. Loch Ness was formed by a tectonic rift that was later deepened by glaciers, and Loch Morar is glacial in origin and separated from the ocean only by a narrow barrier of glacial debris.

Loch Ness drains 10 times more country and is affected a great deal more by human activities than Loch Morar. It is surrounded by 20 times more people, 100 times more roads, and collects the runoff from at least eight sewage treatment plants. Yet in spite of the inflow of potassium and nitrogen, Ness's peaty-brown waters are relatively unproductive.

Loch Morar is even less productive. Its water is about twice as clear as Loch Ness's, with objects visible more than 30 feet beneath the surface. Neither lake supports much aquatic vegetation, since both shelve steeply near shore and have little shallow water. Cold water, depth, and lack of nutrients combine to limit primary production, with the result that few animals can live in the lakes. They are dominated by aquatic insects, especially stoneflies, and by brown trout, Atlantic salmon, char, sticklebacks, and eels. Morar is inhabited by only 71 known species of animals, including just one species each of flatworm, leech, and mollusk.

Curiously enough, both lochs have their resident monsters. Reported sightings of Nessie go back over 1,000 years and have upstaged the less-publicized monster of Loch Morar. The best argument against the existence of either monster is the impoverishment of the lakes. Creatures of such size would require a great deal of food—whether they're filter-feeders or predators—and most biologists agree that the lakes are not productive enough to support breeding populations of Nessies. One hypothesis about Nessie is that the monster is a plesiosaur that somehow survived while going extinct elsewhere 100 million years ago. But Loch Ness was entirely capped with glacial ice just 10,000 years ago, so it is unlikely a population of relics could have survived there.

SHALLOW LAKES

In many cases shallow lakes are more productive than deep ones. Stirred by waves, their organic material circulates quickly rather than ending up far down on the bottom in deep water where it might never return to the energy cycle. They are also usually warmer than deep lakes, which speeds decomposition and the life cycles of organisms. Primary production is high, which creates a food base for more diverse invertebrates.

Shallow lakes in arid regions occasionally dry up altogether, making life difficult or impossible for fish that breathe by passing water over their gills and extracting dissolved oxygen into blood vessels. Oxygen can be scarce in warm, shallow water, making gills useless. An ancient solution for certain fishes is to breathe through lungs. The air-breathing fishes now found in Australia, South America, and Africa are little changed from fossil lungfish dating back 300 million years. The walking catfish of the genus Clarias, found across much of Africa and Southeast Asia, can obtain oxygen from air by means of a branched structure protruding from its gills. When its habitat dries up or otherwise becomes unsuitable, the catfish walks overland on its fins until it finds a pond or stream to its liking.

Regardless of its size, every lake is constantly modified by wind, current, erosion, and sedimentation. As waves and current batter a shore it erodes, causing silt and other easily transported material to drift downward. Sediments get sorted by size, typically with rock and gravel left near the beach and progressively finer sand and silt building outward from shore and forming a terrace. The terrace, known as a *littoral shelf*, often drops abruptly into deep water in what is popularly known as a "drop-off." Other changes are occurring all the time. Current alters bottom configuration; plants take root and grow and die and add their remains to the sediment load. Nutrient levels increase, altering the types and numbers of aquatic organisms.

The lake I grew up on in northern Michigan is becoming more mature every year. As a child I noticed the way the water changed each season, from blue clarity in the weeks following ice-out in March to a greenish hint of algae as the summer progressed. Older people around the lake said they remembered when the water stayed clear all year and there were no weed beds around the islands. The lake is changing, they said. Only a few of them realized that the fertilizer they scattered on their lawns was one of the agents of change.

In time Long Lake is destined to fill with sediments until even its deepest water grows thick with aquatic plants. In the inevitable process of ecological succession, the lake will become a bog, then a meadow, then a forest. Nobody expects it to happen anytime soon, but someday blueberries will ripen where we now troll for walleyes.

WETLANDS

About 6 percent of the earth's land surface is covered with bogs, marshes, swamps, fens, morasses, moors, mires, and floodplains. For centuries wetlands were considered useless mosquito-breeding wastelands that needed to be filled as soon as possible. It is estimated that more than 50 percent of the wetlands present in the world in the year 1500 have been lost as a result of human activities. Some 90 percent of New Zealand's swamps and bogs have been destroyed since European colonization, and more than 40 percent of the coastal wetlands in Brittany have been filled just since 1960.

In the United States about 360,000 square miles of swamps, bogs, and marshes have been eliminated in the past 500 years. Draining and filling swamps to make way for agriculture and development has virtually been a national mandate. In 1763 George Washington was a founder of a company formed to drain North Carolina's and Virginia's Great Dismal Swamp and convert it to farmland (the plan failed). The Swamp Land Acts of the middle 19th century encouraged and financed drainage programs that destroyed about 100,000 square miles of wetlands. During the 20-year period that ended in 1970—before widespread appeals by environmentalists and sportsmen to halt the destruction—more than 450,000 acres of wetlands each year were drained for agricultural use. Farms have accounted for nearly 90 percent of the wetland loss in recent years in the United States, while urban development accounts for 8 percent.

Wetlands are among the most fertile and productive of all environments. They breed mosquitoes, granted, but they breed countless other animals and plants as well. Coastal wetlands, with their blending of salt and fresh water, are critical spawning grounds for many marine animals, including shrimps and dozens of fish species, and support a diverse biota of plants and animals, from plankton to cypresses, mangroves, and birds. Freshwater wetlands are equally essential to wildlife and in heavily developed regions provide the only remaining habitat for some species. They also serve as buffers against flooding and as natural filter systems that remove pollutants from surface water.

Defined as transitional zones between dry uplands and deep-water habitats, wetlands are found wherever the water table is near the surface and land is covered by less than six feet of water. Among the most familiar of the many types of wetlands are *freshwater marshes*, which are characterized by heavy growths of pickerelweed, cattails, and

emergent plants that grow partly in and partly out of the water. Marshes can be covered with water to several feet deep, and can fluctuate with the seasons. They often begin as shallow lakes that have filled with sediment, or as floodplains, sloughs, and oxbow lakes along rivers. The largest freshwater marsh in the world is the Florida Everglades, which encompass 2,700 square miles of Florida, from Lake Okeechobee to the southern tip of the state. Among the many smaller examples found elsewhere are the "dambos" of southern Africa, used as agricultural and grazing lands by villagers and as habitat for wildlife, and the several million pothole marshes of the North American prairies, which are important nesting sites for waterfowl and other birds.

Swamps and *floodplain forests* differ from marshes in being dominated by shrubs and trees, both deciduous and evergreen. Trees and shrubs can survive in a swamp only if they can tolerate periodic flooding or permanent standing water. In North America, the most successful swamp trees include red and silver maples, willows, cottonwoods, cedars, firs, spruces, sweet gum, tupelos, and bald cypress. The world's largest floodplain forests are in the Congo Basin in Africa and the Amazon floodplain in South America.

In formerly glaciated regions, most lakes are destined to become *bogs*. These low, spongy wetlands are composed of a substrate of peat instead of the mineral soils of swamps and marshes. They usually form only where prolific growths of plants live in poorly drained soil with abundant water, high acidity, low temperature, few nutrients, and not much dissolved oxygen. Those conditions make it difficult for bacteria to thrive, thus slowing decomposition of plants as they die and allowing them to accumulate in peat deposits that can become 30 or 40 feet thick. Bogs usually support only acid-tolerant plants such as sphagnum moss, cotton grass, certain sedges and rushes, larch, tamarack, black spruce, and insectivorous plants like the sundew and pitcher plant.

Estuaries, mangroves, and *tidal flats* are characterized by tidal cycles and a mixing of salt and fresh water. Inshore marine waters and estuaries are among the most fertile of all aquatic habitats. Estuaries typically have zones of tidal flats (sand or mud), salt marshes, and rocky outcroppings, all of which support communities of plants and animals. Mangroves are dense-growing and diverse coastal woodlands composed of as many as 80 species of plants.

LIFE IN A FROZEN LAKE

On the surface, a frozen lake appears lifeless and unpromising, as featureless as an unadorned wall. If it is covered with fresh snow it can look like a frozen desert.

But beneath the snow and ice there is living water. Cut a hole large enough to look through, stretch out on the ice and cover your head with a coat, and you can look into the strange, greenish world below. When your eyes adjust you might see the finely etched leaves of aquatic plants on bottom, an occasional small school of perch or shiners. Look more closely and you can see minute organisms drifting like green stars in a constellation, or surging past like tiny frantic oarsmen. Now and then, if you are lucky, you might see a northern pike or a brown trout or a bass.

When ice and snow form a lid over an inland lake, the amount of dissolved oxygen in the water becomes the critical factor determining the life in it. When the water first freezes over, there is plenty of oxygen and it is distributed evenly from top to bottom. But as the ice thickens, life in a lake grows less easy. If the ice remains clear, photosynthesis continues—enough to support colonies of algae on the underside of the ice—but as little as three or four inches of snow can absorb and reflect more than 90 percent of the sunlight. With photosynthesis reduced, oxygen production dwindles. The supply is used up first in the deep water, forcing fish to rise toward the surface, until by late winter it is not unusual to find fish living only in the top few feet of the lake. If the snow cover is heavy and prolonged, the lake can become so depleted of oxygen that entire populations of fish and other organisms die. These "winterkills" are the primary reason that shallow ponds and lakes in the north cannot support fish.

Many strategies have evolved to handle such challenging conditions. Frogs, crayfish, and many insects burrow into the bottom sediment and enter states of diapause, lowering their metabolism and reducing their oxygen requirement to nearly nothing. The more than 400 species of diving beetles living in lakes and ponds in North America are air breathers. They

carry a stock of air with them as they swim and must replenish it periodically by swimming to the surface to poke their abdomens through the surface film. In winter, when the surface is sealed with ice, some species of the beetles respond to the challenge by entering diapause. Others, however, don't give up: They spend the winter breathing from bubbles of air trapped on the underside of the ice.

Of all the tough creatures that survive in extreme environments, few can compare in hardiness to those that live in the frozen lakes of Antarctica. Some Antarctic lakes lie buried beneath 10,000 feet of permanent ice cap and are probably fed by water melting from the base of the ice. Others are exposed to air but remain capped with several feet of ice year-round, or may freeze solid from top to bottom during the winter. A few are salty, fed by inflow from melting glaciers and evaporated by winds that take the water but leave the salts.

Life in the lakes begins with drifting algae and diatoms and with thick mats of blue-green algae. Mosses grow on the bottoms of some of the lakes, even at depths of 60 to 100 feet where no sunlight penetrates. Within the mosses and mats of algae live entire communities of protozoans, rotifers, tardigrades, and nematode worms. The lakes support only a few species, but millions of individuals. Rotifers, for instance, congregate in such numbers that they form dense, red patches on the bottom. They can withstand freezing every night and thawing every morning for months on end, and in experiments have survived temperatures of minus 108 degrees Fahrenheit. Several species of tardigrades can be dried and frozen in the harsh Antarctic wind, then will swim off unharmed when submerged again in water.

A larger inhabitant of Antarctic lakes is a species of fairy shrimp closely related to shrimps that live in temporary pools in deserts. Like the desert shrimps, it has highly resistant eggs that can survive the most adverse of conditions and will hatch only when placed in water. The adult shrimps die when the lake freezes again, but not before laying eggs to ensure the next generation's survival.

Seiches

One evening I was wading in shallow water at the tip of Old Mission Peninsula when I got caught by a squall. I was a quarter mile from shore, walking knee-deep over the gravel bar where East and West Bays join together to form the main body of Grand Traverse Bay. To the north the shores curved out of sight and open Lake Michigan spread across the horizon.

From the west, over the hills of Leelanau County, came a squall of rolling black clouds that charged the five miles across West Bay so quickly that I knew I had no chance to make it to shore. As the lightning approached, I kneeled in the water, trying to make myself as small a target as possible. Then the wind came, pushed ahead of the squall-line like dust ahead of a broom. It flurried the water, raised instant whitecaps, and seemed to shove the whole mass of West Bay's water across the bar into East Bay. For half an hour, as rain fell and lightning flashed, the wind blew strong to the east.

As abruptly as it arrived, the squall passed. The wind fell, as if someone had slammed a window shut, and the waves calmed.

Then I noticed something curious: Gently, almost imperceptibly at first, a current began flowing from the east to the west. It crept past my knees over the bar, forming eddies where stones broke the surface and sending little spirals spinning into West Bay. Gradually the current grew stronger until it was audible, a river of lake water whispering and shushing around me, swirling around rocks and rolling pebbles along the bottom. As darkness fell and I finally waded back to shore, it was still flowing.

It was one of the most dramatic examples I've seen of a seiche. The word, pronounced "saysh," was originally used in the early nineteenth century to describe the peculiar oscillating wave that now and then washes up and down the length of Switzerland's Lake Geneva. It has since been applied to similar waves on lakes all over the world.

Seiches behave somewhat like water in a bathtub. If you move slowly in a tub, the water is scarcely disturbed. But slide abruptly from a sitting to a reclining position, and the displacement of water by your body forces it to rush to the end of the tub, where it reflects and sloshes back several times before subsiding. A similar sloshing occurs in bays, inlets, gulfs, and lakes when strong wind or sudden changes in atmospheric pressure create enough force to displace water. In bays open to the ocean, seiching sometimes follows the arrival of trains of long-period swells and tsunamis. One common result of seiches in harbors is the "surging" of moored boats. As the long waves of the seiche reflect back and forth across the harbor, the boats strain at their mooring lines or spin slowly around their anchors.

In Lake Geneva, seiches occur frequently and are often about three feet high. The largest in the Great Lakes usually occur in Lake Erie, where the record seiche occurred in January 1942 when water stood more than 13 feet higher on the shore of Buffalo than at Toledo. In Lake Erie seiches routinely raise and lower the water levels in harbors, occasionally stranding ships or causing them to run aground, and occur frequently enough that lake freighters schedule their docking and departing times to avoid them. Usually they are quite harmless.

One that was not harmless occurred along the south end of Lake Michigan on June 26, 1954. Early that morning the lake was unusually calm, but about eight o'clock a thunderstorm with winds clocked at more than 50 miles per hour blew in from the northwest. At Michigan City, Indiana, 100 people fishing on a pier scrambled to get out of the way as a two-foot-high surge of water washed over the pier. This was the *incident wave* of the seiche, plowed ahead by the approaching storm. Nobody on the pier was injured or much concerned. But, unknown to them, as the storm passed overhead and beyond the lake the water it had pushed to shore rebounded and returned to the northwest. As this *reflected wave* approached Chicago the bottom of the lake created a shoaling effect, forcing the wave to drag, slowing it and compressing much of its energy into a high crest, in a process identical to what happens when a tsunami nears shore. By the time the wave reached the Chicago waterfront it was 10 feet high. People on jetties and piers along the waterfront were caught off guard. Dozens were washed into the lake. Eight drowned.

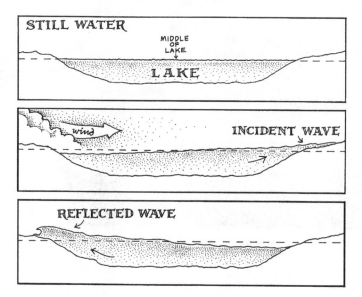

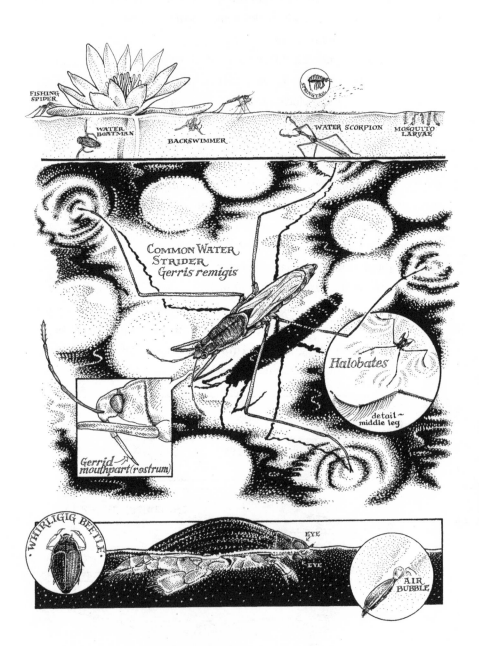

FISHING SPIDER

SPRINGTAIL

WATER BOATMAN

BACKSWIMMER

WATER SCORPION

MOSQUITO LARVAE

COMMON WATER STRIDER
Gerris remigis

Halobates

detail ~ middle leg

Gerrid mouthpart (rostrum)

WHIRLIGIG BEETLE

EYE

EYE

AIR BUBBLE

14
LIFE ON THE SURFACE

What luxury it was for my brother and me to rise early on summer mornings when the grass was still drenched with dew and run outside in sneakers and cutoffs and ratty T-shirts to spend the day exploring ponds. Our sneakers and cutoffs were seldom dry those summers. Mom worried about mildew.

Rick and I spent hours every day at two small ponds in the woods across the road from our house. One was a seasonal pond that gathered every spring in a depression beneath the maples and was usually dried to scaly mud by the middle of the summer. The other was lily-fringed and big enough to grow bass and bluegills but so shallow that the fish were sometimes killed during hard winters when ice and snow blocked the sunlight and the water grew starved of oxygen.

We collected pond water in specimen bottles and took it home to our makeshift laboratory in the basement to study under a microscope. Every drop was so filled with bizarre wriggling, spinning, and pulsating organisms that examining them through a lens was like glimpsing communities on another planet. It was a little disconcerting. Life seemed to be busting out all over the place. We watched paramecia swim by waving their hundreds of hairlike cilia and amoebas stream liquidly across the slide by extending their false feet ahead and pulling themselves behind. The water contained far too many protozoans, rotifers, copepods, water fleas, and tardigrades for us to identify.

Larger creatures were less baffling. We captured and took home frogs, toads, water snakes, mudpuppies, and snapping turtles. We netted damselflies and dragonflies, and herded schools of shiners and darters into the shallows and trapped them in jars.

Just the insects in a pond are worth studying for a summer. A backswimmer floats tail-up near the surface, breathing from a bubble of air embraced against its abdomen. It swims underwater, upside down on a convex back, its body covered with water-repellent hairs that trap air and give the insect buoyancy and a silvery appearance. It uses its hind legs like oars, flat and edged with bristles, to backstroke rapidly underwater.

Closely related is the water boatman, a bug similar in many ways to the backswimmer except that it swims belly down, propelling itself with broad strokes of the legs positioned near its head. Like the backswimmer, it breathes air that it carries like scuba equipment, but unlike the carnivorous backswimmer, which feeds by grasping and sucking the body juices from small insects, tadpoles, or very small fish, the water boatman is a plant eater.

Some aquatic insects are such efficient predators they can be terrifying. The giant water bug, for instance, which measures up to three inches long, can impale a frog with its switchblade forelimbs and inject it with a chemical through its proboscis. Within a few minutes the chemical dissolves all the tissues inside the frog's skin, allowing the beetle to suck it back through the proboscis. Water scorpions slowly prowl the bottom, grasping their prey with forelegs as quick as a boxer's jabs and sucking their insides much like the giant water bug. When a water scorpion needs air it climbs a plant stalk until it is near the surface then thrusts a pair of appendages on its posterior into the air to serve as snorkels.

There are also diving beetles, members of a large family (Dytiscidae) found all over the world and represented by more than 400 species in North America alone. They collect air beneath their forewings, then swim rapidly by stroking their flattened hind legs in simultaneous strokes—

unlike the alternating strokes of most aquatic beetles—until the air in their chamber and a small additional amount they can absorb from the water is used up and they must go to the surface again. Most dytiscids are protected from predators by defensive chemicals secreted from a gland at the tip of the abdomen, including benzoic acid and benzaldehyde, and (uncommon among insects) such steroids as testosterone and estradiol. The steroids cause any fish, amphibian, and mammal reckless enough to swallow a diving beetle to become nauseated and vomit.

In our random way, Rick and I collected specimens of nearly every creature that lived within those Michigan waters. But life on the surface eluded us. While sitting on a log or stretched out in the grass beside the pond, we would watch water striders and whirligig beetles darting and spinning across the surface and wonder why bluegills never rose to feed on them the way they rose to mayflies and caddisflies and dragonflies. We wondered what those maddeningly evasive insects ate, how they reproduced, whether they could dive beneath the surface, whether they would fly away if we tossed one into the air. But they were too quick for us. We never added them to our collection.

Years later we would learn that all the floating, swimming, skating, and jumping creatures that live in that thin habitat at the top of the water are members of a community of animals and plants known as the *neuston*. The only requirement for membership in the neuston is the ability to live in or on the surface film. Mosquito larvae hang from it like bats from the roof of a cave; small snails float there on bubbles of air; springtails wriggle in the film and catapult themselves into the air with a flick of their strong tails; fish spiders scoot across the surface, buoyed by waterproof hairs, then dive to hunt for small aquatic animals.

To understand why an insect or spider can walk on the surface of a pond, or why, for that matter, you can float a needle on water, it helps to remember that molecules of water always attract one another, which is why water tends to stay in a pond or ocean instead of evaporating instantly, and why water on a windshield collects tirelessly into drops. Beneath the surface the molecules are attracted to each other equally in all directions. At the interface between water and atmosphere, however, the attraction of the water molecules to one another is greater than their attraction to the air molecules above. This disparity causes the molecules on the surface to cling together in a taut membrane. We hardly notice the membrane, but to a small organism it can be all but impenetrable, allowing them to walk on it like a floor or hang from it like a ceiling or be trapped within it like quicksand.

What we so casually call bugs, those specimens of the vast, varied, and to many people repulsive world of insects, arachnids, and other small crawling and flying creatures, have a more specific meaning among scientists. Bugs, to an entomologist, are insects belonging to the order Hemiptera and are characterized by having piercing-sucking mouthparts and wings that are leathery at the base and membranous at the tip. They include such land dwellers as lace bugs, assassin bugs, stinkbugs, leaf bugs, and bedbugs, and such aquatic and semiaquatic bugs as water striders, water scorpions, toad bugs, shore bugs, water boatmen, and backswimmers. Anyone who has spent time watching the surface of ponds is probably familiar with several specialized members of the order of bugs. Forget bedbugs and stinkbugs: Water striders are the darlings of the order.

Some 1,500 species of surface-dwelling hemipterans can be found on every continent except Antarctica. The best known of them are members of the family Gerridae, which in North America is represented by nearly 50 species, known collectively as water striders, pond skaters, or—because of their ability to walk on water—Jesus bugs.

Water striders do most of their remarkably deft travel over the surface film with the help of enlarged second and third legs. The front two legs are smaller, are held folded under the head, and are equipped with sharp claws used for grabbing prey. The hind four legs are long and thin and lined, on their lower halves, with waterproof hairs that are all that make contact with the water's surface. The hairs are so minute that the attraction between water molecules remains stronger than the attraction between the water and the hairs, keeping the surface film intact. The insects use their legs like long oars to propel them across the surface, the middle two for propulsion, the hind two for steering.

Their mobility makes water striders masters of the surface film. Any small insect that falls to the water can become a victim of those grasping legs and sharp mouthparts. Aquatic insects that must come to the surface for air or to metamorphose are also frequent prey; the moment they break the surface, transmitting ripples across the water, the striders close in.

That sensitivity to surface vibrations shows up in courtship as well. When a male seeks a mate he attaches himself to something solid—a plant or a piece of floating debris—and raps the water with his rowing legs ten to thirty times each second. An interested female responds by sending out ripple signals of her own at a slower rate.

In northern regions, where ice-up makes life on the surface treacherous, water striders respond in an unusual way to the change of seasons. Several generations of striders mate, lay eggs, hatch, and die during the warm months, each generation likely staying on the same body of water. But those that emerge late in the year, when cold weather diminishes food supplies and threatens to freeze the water, sometimes have a choice. In a condition known by biologists as alary (or wing) polymorphism, some of the individuals within each population are born with wings while others are not. The wingless adults spend the winter burrowed beneath leaf litter, logs, or rocks within crawling distance of the water. Those with wings travel much farther from the water before burrowing for the winter.

The absence or presence of wings among individual water bugs has long been one of the puzzling mysteries about the aquatic and semiaquatic hemiptera. One theory proposes that population density triggers some individuals to be born with wings, allowing them to fly off and colonize new waters. Other theories assume that random numbers of individuals are born with wings for the same purpose.

Some wingless species of striders, members of the genus *Halobates*, are ocean dwellers. About five species, commonly known as sea skaters, are among the only insects to have colonized the open ocean. They've been discovered skating up and down swells as far as 200 miles from shore, engaged in a struggle for survival that seems daunting to us land lovers. *Halobates* are solitary hunters of small creatures that have become caught in the surface film, feeding on them in the same manner as the freshwater gerrids. They mate and lay their eggs on bits of flotsam and are at the mercy always of wind and current. *Halobates*, not surprisingly perhaps, are the least-studied of any of the water striders.

On the tiny ponds near my home I watched water striders for hours at a time. They stood poised and motionless on their delicate legs, now and then turning abruptly, first one way and then another, shifting directions like an oarsman in a boat who pulls back with one oar and pushes forward with the other. Frequently some stimulus set the insects in motion and they scattered at high speed, like frantic skaters from a rink. I stunned a mosquito and flicked it into the water. Invisible ripples telegraphed across the surface and a strider charged and straddled the mosquito possessively, like a hawk hooding a field mouse. Something busy and horrible then occurred. Contact of some sort was made between the mouth of the strider and the inert body of the mosquito, but it took place at a scale too small for my eyes. I assumed the mosquito was being devoured but I could see no munching mouthparts, no tearing or ripping. The strider lost interest after a few minutes and skated away, the mosquito left apparently unchanged.

Years later I learned that the mosquito had been changed in a dreadful way. In the library, reading about insect behavior with the guilty fascination I would feel reading biographies of serial killers, I learned that water striders were more gruesome predators than I had imagined. The mosquitoes had not been devoured so much as converted into soup and slurped. A strider, holding its prey with the claws high on its front legs, would pierce it with its mouthpart and release an enzyme that liquefied all the internal body parts. It then sucked the liquid out. The mosquito that was left behind was just a shell of its former self.

Gruesome diet aside, water striders are among the most graceful and intriguing inhabitants of ponds, lakes, oceans, and slow rivers. That they are surface dwellers makes them special. We may be fascinated with subsurface life, but the world beneath the surface will always be foreign and dangerous to us. The surface is where we live. It is where we see the reflection of autumn leaves, spangles of sunlight, and the wavering trails of moonlight. It is the domain of Narcissus, who stared so long at his own reflection that he fell in love with it, and of the photographer Eliot Porter, whose images of pond surface and creek pool illustrate so aptly the wildness of the world as described by writers like Thoreau and Leopold. We dive into water, wishing for the grace of otters and penguins, but must settle for a clumsy splashing wallow. What we want, really, is to skate as freely and gracefully across the surface of our planet as a water strider skates across the surface of his pond.

THE AMAZING WHIRLIGIG

In many places, water striders must share habitat with another denizen of the surface: whirligig beetles. These strange beetles, the only members of the order Coleoptera to be admitted to the neuston community, are splendidly equipped for the two-dimensional netherworld where water meets air. A whirligig, first of all, is aptly named, even in Latin: The family name, Gyrinidae, is from the root *gyr*, to spin or whirl. To escape enemies and hunt prey it whirls across the surface in tight circles at speeds up to two knots, the fastest of any aquatic insect. It achieves those speeds by kicking oarlike hind legs 60 times per second, faster than the wingbeats of some hummingbirds. Abrupt changes of direction are made possible by a flexible segment at the end of the abdomen that acts like a rudder.

Whirligigs are black, shiny, and streamlined, like 1949 Buick sedans. They're often found resting by the hundreds in tightly bunched masses on

WHIRLIGIG
LARVA

calm water in lakes, ponds, and slow rivers. But disturb them with a canoe paddle or a rock and they speed off in all directions.

The whirligig beetle is perfectly at home in the surface film. Its teardrop-shaped body is lubricated with oils to repel water, and it is equipped with two sets of eyes, one below the waterline for seeing underwater, the other above the waterline for scanning the surface and sky. If pursued by a predator it can dive and swim rapidly underwater or can launch into flight and travel up to two miles by wing. When it dives underwater it carries a supply of air stored in a bubble beneath its elytron (the hardened forewing covering its back) and clings to a plant stem or other anchor. When it releases its hold it pops to the surface like a cork.

Fish, frogs, and other predators learn quickly that whirligigs make poor meals. Besides being extremely adept at escape, when threatened they emit noxious chemicals that make them virtually inedible. In North America, those whirligigs in the genus *Gyrinus* (measuring about a quarter-inch long or smaller) give off an offensive smell powerful enough to be detected by the human nose. Those species in the genus *Dineutus* (which measure up to half an inch long) give off a more appealing scent that many people say reminds them of apples, accounting for their common name in some regions: "apple bugs."

15

SALT LAKES AND
SEA MONKEYS

Superman and Wonder Woman can't hold a candle to sea monkeys. That's clear to me now, but when I was 12 and studying comic-book advertisements for pet chameleons, flesh-eating plants, pen-size spy telescopes, and secret Eastern techniques for transforming my bare hands into lethal weapons, I was skeptical. The ads claimed that for a dollar I could receive a packet of mysterious eggs that would hatch into creatures guaranteed to fill my life with wonder and enchantment. But I was a shrewd kid, not easily duped. I spent my money instead on hypnotism discs and X-ray glasses.

If I had taken a chance on sea monkeys, however, I might have learned that they were not some bizarre species of aquatic primate, but brine shrimp, half-inch-long crustaceans native to certain saline lakes, the most famous being Utah's Great Salt Lake. If I had been less of a skeptic, my fishbowl might have been home to some of the world's toughest survivors.

Great Salt Lake is only one of countless salt lakes around the world, many of them habitats so harsh even most bacteria find them impossible to live in. They are found on every continent, including Antarctica, where warm saline water at the bottom of Lake Vanda is covered with a permanent cap of ice a dozen feet thick. Although there is much variation among salt lakes, all share a single characteristic: They have no outlets. Because they are located below sea level or in basins surrounded on all sides by hills, the water entering them by rain, river, or runoff stays put. Or, rather, the solvent material in the water stays. The water itself evaporates, leaving behind a gradually increasing accumulation of chemicals and minerals. When the accumulation exceeds about 3 percent—nearly equaling the 3.5 percent average salinity of the oceans—the body of water is considered a salt lake.

The chemicals and minerals in a salt lake vary according to the composition of the soil and rocks in the drainage system feeding it. Each lake is distinct, a chemical fingerprint of the land around it. The salinity in Great Salt Lake is derived primarily from sodium chloride— ordinary table salt—and sodium sulfate. The soda lakes of East Africa's Rift Valley contain dissolved carbonates. California's Mono Lake, shrinking rapidly as the rivers that feed it are diverted to supply the water needs of thirsty Los Angeles, contains an ever-thickening mixture of sodium chloride and sodium carbonate. The water of the Dead Sea, the saltiest lake of all, contains sodium chloride, potash, magnesium, bromine, and other minerals.

Great Salt Lake is no pond—it spreads across 2,000 square miles of sun-baked desert in northwest Utah—but it is just a remnant of a much larger body of water known as Lake Bonneville. Fifteen thousand years ago Lake Bonneville covered 20,000 square miles of Utah, Nevada, and Idaho and drained north to the Pacific through what are now the Snake and Columbia Rivers. As the level of the ancient lake declined, salt accumulated. Some of it was deposited on its former bed as hard salt pans, part of which is now known as the Bonneville Salt Flats, and a lot of it ended up in Great Salt Lake. Geologists estimate that each year more than three million tons of salt finds its way into the lake.

That anything can live there is a wonder. The shores of the lake are rimmed and the bottom coated with compacted, cementlike layers of

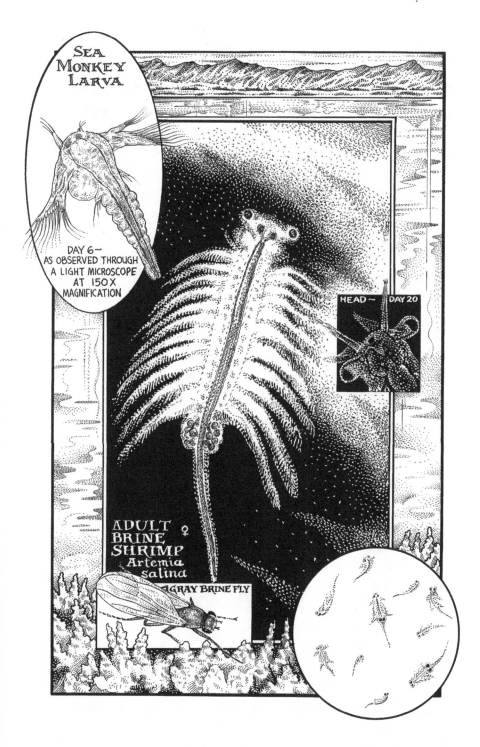

SEA MONKEY LARVA

DAY 6~
AS OBSERVED THROUGH
A LIGHT MICROSCOPE
AT 150X
MAGNIFICATION

HEAD~ DAY 20

ADULT ♀
BRINE
SHRIMP
Artemia
salina

GRAY BRINE FLY

crystallized salt, and the water itself is so dense with salts that humans find it virtually impossible to swim beneath the surface. The deepest portion of the lake (nowhere more than 50 feet, with an average of only about 13 feet) are so salty they contain no oxygen. Even the less salty surface water holds little oxygen because of warm temperatures and high altitude. Freshwater of the same temperature (about 77 to 86 degrees Fahrenheit) can hold at least eight times more dissolved oxygen than the water of Great Salt Lake. In the summer, the shore of the highly saline north arm of the lake is often fringed with a reddish-colored scum of bacteria and algae trapped beneath layers of crystallized salt. The algae receive enough sunlight to photosynthesize, producing oxygen that accumulates beneath the salt. But even that bounty of oxygen is unavailable to organisms in the lake. Under pressure, the salt rises in small domes of gas, splitting open at the surface and releasing bursts of pure oxygen into the air. Very little of it dissolves in the water.

Life in Great Salt Lake faces the further challenge of fluctuating water level. Beginning in 1847, when Mormons first settled on the banks of the lake's primary tributary and named it the Jordan River, they discovered that Great Salt Lake rose and fell several feet each year, from a high in winter and spring to a low in summer. Its fluctuations were even greater in the long term. In 1875 the lake stood at 4,234 feet above sea level. By 1963, after a long series of ups and downs, it fell to a record low of 4,214 feet. That 20-foot drop caused salt concentrations in the lake to reach 10 times the concentration of ocean water.

A number of organisms have successfully colonized Great Salt Lake's inhospitable environment, evolving into a simple aquatic food chain with bacteria and algae at the bottom and one species of brine shrimp and the larvae of the gray brine fly at the top. Other inhabitants, including a protozoan, a fungus, and a few viruses, live in the less salty regions near the mouths of tributary rivers, but in the main body of the lake there are only bacteria, algae, shrimp, and brine fly.

Though this exclusive community is simple, it is astonishingly crowded. A rule of thumb in salt lakes (as in many extreme environments) is that the number of species is low and the number of individuals is high. When the population of brine shrimp is at its peak, the numbers are too vast for comprehension. In 1965, during the height of the sea-monkey craze, more than 77 tons of the tiny eggs were harvested from the lake. Likewise, brine flies hatch from the water in such quantities that they sometimes cover it with a seething black quilt of insects. Such abundance does not go unnoticed: The shrimp and flies of Great Salt Lake attract hundreds of thousands of shorebirds and waterfowl.

Birds are also drawn in vast numbers to the fertile but caustic waters of Africa's soda lakes, which, though considered saline, contain dissolved sodium carbonate rather than sodium chloride. The soda lakes are clustered in the Western Rift and Eastern Rift Valleys, and vary greatly in size and salinity, from tiny ponds in ancient volcanic craters, to lakes hundreds of square miles in area and reaching depths of more than 825 feet. Two of the lakes with the widest reputations are Nakuru, in Kenya, and Natron, just across the border in northern Tanzania.

Like Great Salt Lake, Lake Nakuru is inhabited by a simple community containing a few species that thrive in very large numbers. The basis of the food chain is a blue-green alga, which is fed upon by three consumers: a copepod, the soda lake fish, and the lesser flamingo. A few species of rotifers, four species of water boatman, and the larvae of two species of midges round out the community.

Lake Natron has a nasty reputation for being highly corrosive, its soda content so high that contact with bare skin causes dangerous burns. It is bounded by mudflats covered with an alkali crust that may or may not hold the weight of a human being, and is spotted with icelike rafts of soda kept afloat by accumulated gases. Thermal springs and geysers provide a steady flow of hot water and soda into the lake, and daytime temperatures of up to 150 degrees Fahrenheit evaporate water faster than it can be replenished. The biotic community consists of algae, insect larvae, brine shrimp, and a bacteria colored with a bright-red pigment that is one of the few substances besides chlorophyll able to transform sunlight into food.

The most spectacular inhabitants of both lakes are flamingos. The lesser flamingo, *Phoeniconaias minor*, is especially prevalent, its numbers estimated some years at 1.5 million, all of them feeding in constantly moving ranks of pink around the shores of the lakes. Each bird feeds by inverting its head and sucking water through its peculiar, downturned beak, which is equipped with filter platelets that strain algae and other small organisms from the water. It's hard work: Each flamingo strains about 8 gallons of water an hour, 12 hours a day. If algae are extremely abundant, that adds up to about 2.5 ounces of food per day per bird. When lake levels rise, however, the algae disperse or are replaced by smaller species that thrive in less saline water, forcing the flamingos to migrate. They move up and down the Rift Valley, visiting various lakes strung out in a system of soda lakes about 300 miles long. The greater flamingo, *Phoenicopterus ruber*, is present in smaller numbers and feeds by straining the water for larger organisms, including copepods and midge larvae.

The five species (and three subspecies) of flamingos in the world are the only birds specialized for feeding in saline and soda lakes. The lesser flamingo is the most abundant and lives mostly in the Rift Valley. The greater flamingo and its subspecies are more widespread, with breeding populations in East Africa, India, Iran, Tunisia, the salt lagoons along the southern coast of France, and on several Caribbean islands, the Yucatan coast of Mexico, and the Galapagos Islands. Of the three species found only in South America, the Chilean flamingo is the most abundant and widespread. It and the Andean and James' flamingos are found at high elevations, often more than 13,200 feet above sea level, in salt lakes in the mountains of northern Chile. The flamingos weather the extreme cold of winter by gathering in warm springs near the salt lakes.

Another remarkable inhabitant of Africa's soda lakes is *Oreochromis alcalicus grahami*, the soda lake fish. This cichlid, measuring four to ten inches in length, is a resident of some of the most difficult habitats of any known fish: hot alkaline springs located near the shores of soda lakes. The fish lives in water as hot as 98 to113 degrees Fahrenheit and feeds on blue-green algae adhering to rocks in the springs. The algae are extremely heat-tolerant and thrive in the hottest portions of the springs, luring the fish to the edge of disaster, where they leave distinct browse lines at the point where temperature becomes intolerable. Like many other members of the large and diverse family of cichlids found throughout Africa, Asia, South America, and North America, the soda lake fish broods its young in its mouth. The female carries her eggs in that safe haven, keeping them irrigated with a constant flow of water through her gills, then releases the young when they hatch. Unlike other cichlids, however, the soda lake fish does not carry her young around after they've hatched, probably because no predators can live in its hot-water environment so there is little risk in letting them swim unattended.

Among the most successful of the organisms to adapt to lakes of brine are the brine shrimp, members of an order of crustaceans, Anostraca, sometimes referred to as the fairy shrimps. The order is celebrated for

its ability to survive extreme conditions, thriving in salt lakes, in temporary desert puddles, and in lakes that are frozen most of the year. Unprotected by a carapace and unable to swim at the lightning speed of many of their marine relatives, they rely on their habitats to keep predators away. Relatives of Great Salt Lake's *Artemia salina* live in salt lakes all over the world. Typically, they feed by swimming on their backs, rhythmically beating the water with their appendages, filtering algae and bacteria from the water. They are sometimes pink in color, due to hemoglobin, which assists them in extracting oxygen from oxygen-poor water.

When it comes to reproduction, brine shrimp hedge their bets by producing two kinds of eggs: a thin-skinned variety that clump together in egg sacs in the water and hatch after five or six days, and thick-skinned eggs encased in hard shells, which sink to the bottom and enter a state of restful diapause. Washed up on shore or left stranded as the lake evaporates, the hard-shelled eggs dry out and become tougher than the toughest superhero, able to survive temperatures as low as minus 310 degrees Fahrenheit and as high as 212 degrees. Even after many years of desiccation the eggs will hatch soon after being submerged in water.

Not long ago, Glenn Wolff and I received two packages of brine shrimp eggs in the mail and emptied them into jars of specially prepared saline solution. Our "Sea-Monkeys"—now a registered trademark—arrived in luridly illustrated packages designed for novelty stores. The front of each package depicts a family of cartoon characters sporting buck teeth, knobby heads, and mermaid tails, all grinning like mad and frolicking in front of a submerged castle. "Illustration is fanciful," warns the fine print at the bottom.

Two days later we could see minute pale dots lurching through the water. After four days, they had grown large enough to be clearly identified as brine shrimp larvae. At six days Glenn and I got together in his basement studio, isolated one of the larvae with an eyedropper, deposited it onto a slide, and placed it under a microscope. Magnified 150 times, the shrimp filled the lens and we could see its parts in vivid detail: a feathery and gently pulsating leg, a translucent egg sac, a digestive system flowing with bubbling fluids.

Of course, there was nothing simian about it. Whoever thought of calling it a sea monkey was a marketer not a naturalist. But the old comic-book promises weren't completely empty. Glenn and I took turns looking through the microscope at the brine shrimp, amazed that such a small animal could have such a powerful will to live. Glenn examined it longer than I did. For nearly 15 minutes he sat staring into the microscope. When he finally looked up, his eyes were shining and he was grinning the kind of grin you often see on a 12-year-old boy.

THE HOLIEST SALT LAKE

Though it is not a sea and it's not quite dead, for many people the Dead Sea is the archetype of water made lifeless by too much salt. Thirty miles long and averaging almost 10 miles wide, it is located on the border between Israel and Jordan, in the lowest spot on the surface of the earth, its surface 1,400 feet below sea level and its bottom reaching another 1,086 feet deeper. A few million years ago, when the surfaces of all the oceans were considerably higher than they are today, the Dead Sea was connected to the Mediterranean. But as ice grew around the polar caps, ocean waters receded, landlocking the Dead Sea and leaving it surrounded with deep beds of salt. Since then those ancient deposits have been slowly dissolving and draining into the Dead Sea by way of the Jordan River and the hot, saline springs along the shore of the lake. Except for algae and a rugged species of bacteria so salt-dependent that it dies in water less than three times as salty as the oceans, the sea lives up to its name.

The Dead Sea is the greatest of salt lakes in salt content, which is about 10 times higher than the oceans. Unlike the oceans, where the predominant dissolved salt is sodium chloride, the Dead Sea contains a high percentage of foul-tasting magnesium chloride. Its water is more than one-third composed of salts, making it so dense that waves don't rise in it, the surface can scarcely be ruffled by wind, and swimmers find it nearly impossible to sink beneath its surface. It contains the absolute maximum amount of salt and other minerals possible; not a teaspoon more will dissolve in it. As a result, salt is constantly crystallizing on the bottom and forming elaborate clumps, shelves, and pinnacles of pinkish-colored salt around its shores.

Fed primarily by the River Jordan, where Jesus was baptized, and bounded by the hills of Qumran, where the Dead Sea Scrolls were discovered, the region surrounding the Dead Sea is rich in biblical history but poor in rainfall. It receives an average of less than 4 inches of rain each year, while in the same span of time about 80 inches evaporates from the lake. More than half the flow of the Jordan is diverted for human needs before it reaches the Dead Sea and what is left is not sufficient to replace the amount that evaporates. So the Dead Sea is shrinking. Between 1960 and 2014 its surface dropped more than 80 feet and it shrank more than a third in surface area.

A plan approved in 2013 could save the lake, if Israel and Jordan can get along well enough and long enough to implement it. The plan calls to transfuse the Dead Sea with super-salty water piped 100 miles from a desalination plant on the shore of the Red Sea.

THE DEATH OF A SALT LAKE

Some environmentalists consider what happened to Russia's Aral Sea the greatest ecological disaster of the twentieth century. That large saline lake, once covering an area of 25,000 square miles—larger than all the North American Great Lakes except Superior—is located in a warm, arid region that in the days of the Soviet Union supplied more than a third of the nation's fruits, vegetables, and rice and almost the entire cotton harvest of a country that led the world in cotton production.

But such productivity was purchased at a steep price. Vast collective farms laid out across hundreds of miles of near-desert terrain could not keep up with state-mandated production quotas unless they were constantly irrigated. The only dependable sources of water in the region were two rivers, the Amu Darya and the Syr Darya—the primary tributaries of the Aral Sea. Irrigation claimed so much of both rivers' water that for more than twenty years not a drop of the Syr Darya reached the Aral Sea and only a trickle from the other river reached it. The lake was strangled. The flow of water was so diminished that in 1986, a dry year, none reached the Aral Sea at all. Between 1960 and 1989 its surface fell 46 feet, the area it covered was reduced by nearly half, and salinity tripled. By 2009, it had shrunk even further, to 10 percent of its original size, splitting the water into three separate, shallow lakes. Fishing villages previously located on the shore were now stranded 20 or more miles from it. Water that had for thousands of years supported varied and abundant life, including 24 species of native fish, became too salty to support any fish at all. The mudflats left behind as the water receded are now so encrusted with salts and the residue of massive amounts of fertilizers, pesticides, and herbicides dumped on the adjacent fields that the wind picks them up in swirling clouds of toxic dust and scatters them across the surrounding region. The chemicals have found their way as well into the groundwater supply, poisoning the region's last reserves of drinking water. Without the lake's moderating effect, even the climate of the region has been altered. By some estimates, the growing season of the farmlands north of the lake has been diminished by 10 days per year.

Recent efforts to restore the lake have been somewhat successful. Since 2008, a dam project has raised the level of the North Aral Sea by 40 feet, causing salinity to be reduced and revitalizing a limited commercial fishery.

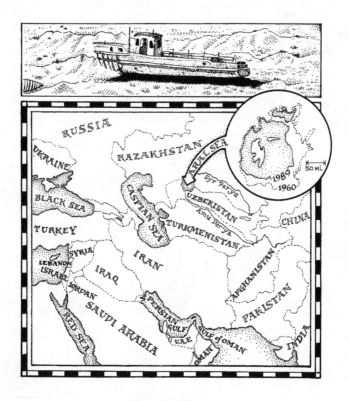

PANGAEA

MARE INTERNUM

COLUMBUS
COOK
MAURY

16
MOTHER OCEAN

Two hundred million years ago there was but one ocean, a vast, globe-girdling sea surrounding a huddled supercontinent called Pangaea. As Pangaea broke up and the continents drifted apart, the land divided the water into parts.

Now all the oceans and seas are separate and named, but they are no more disconnected than fingers on a hand. There are differences among them but they share the same water and are subject to the same laws of wind, current, and tide. Together they still form one planet-embracing ocean.

The Romans called the chain of seas within their empire *mare nostrum* or *mare internum*—"our sea," or "the sea that is known to us." Solinus, a Roman geographer of the third century A.D., was among the first to label those waters "mediterranean," or "sea in the middle of the earth." From classical antiquity to the Middle Ages there was a big difference between a sea and an ocean. There was only one Ocean, known in English as the Ocean Sea, *mare oceanum*, which, like the *Oceanus* of Greek mythology, surrounded the Mediterranean, the Aegean, and the other seas in a boundless, universal body. Whatever lay past it was unknown and unknowable, a realm beyond the reach of living men.

Though there was just one ocean, a seasoned mariner was likely to know seven seas. To "sail the seven seas" now refers generically to all the world's oceans and seas, but it was originally used by ancient seafarers to mean the Mediterranean, Adriatic, Caspian, Red, Black, Persian Gulf, and Indian Ocean. The term fell out of use during the Age of Discovery, when it became apparent that the world contained a lot more water than had ever been imagined, but it returned to common usage after the publication in 1896 of Rudyard Kipling's book of poems, *The Seven Seas*. Kipling was so popular that people were determined to divide the waters of the earth into seven parts, even if it had to be done rather artificially. It was decided that the new seven seas should be the North Atlantic, South Atlantic, North Pacific, South Pacific, Indian, Arctic, and Antarctic.

When Columbus set sail for the Indies in 1492, most Europeans clung dogmatically to the notion that the earth was six-sevenths covered with a continuous landmass that included Europe, Africa, and Asia, all surrounded by a one-seventh proportion of water. The belief originated from the words of Esdras in the Apocrypha, "six parts hast thou dried up," and was sustained by the assumption that God would waste little of the surface of the world on an element that could not nurture the bodies and souls of His highest creations. More water than land was unthinkable. It implied disorder in the Divine plan.

Even armed with the faith that he was crossing a small ocean, Columbus's voyage required great courage. It had taken decades and the cumulative plodding efforts of dozens of expeditions for the Portuguese to explore the western edge of Africa. Coasting was risky; setting off across an unknown ocean was considered suicidal. Columbus read widely in his attempts to learn the size of the earth and its oceans. One source he seems to have referred to a lot (based on his many margin notes) was a geography of the world titled *Imago mundi*, written about 1410 by a French astrologer, Pierre d'Ailly. One passage in particular gave Columbus comfort: "For, according to the philosophers and Pliny, the ocean which stretches between the extremity of further Spain (that is, Morocco) and the eastern edge of India is of no great width. For it is evident that this sea is navigable in a very few days if the wind be fair..."

Columbus wisely chose not to sail west from Spain, which would have taken him into the face of predominate westerly winds. Instead, he first sailed seven days south to the Canary Islands, then turned west, pushed by the trade winds. Columbus's idea, borrowed from Marco Polo's account of his journeys, was that the Canaries were located on the same latitude as Japan and that sailing west would take him straight there. As it turned out, a continent stood in the way, but the winds were favorable. For the next 500 years ships would follow those same northeasterly trade winds to the New World, then return to the Old World on the westerlies of the North Atlantic. It is presumptuous to say that Columbus discovered the New World, since civilizations had thrived there for thousands of years, but he was certainly the discoverer of the winds that made voyages there and back convenient.

*

It is probably not possible for us to imagine the terror the oceans once inspired in people. For us the rhythmic boom and hiss of surf is soothing, but to people centuries ago the sounds of the ocean were more likely to be a reminder of imminent punishment and annihilation. Evidence of sinister intentions seemed everywhere. Storms swallowed ships and sharks devoured their crews. Waves pulverized coastlines and cast up decaying carcasses that hinted at new and unknown terrors in the deep.

The Bible—especially Genesis, the Psalms, and the book of Job—had a powerful influence on how the oceans were perceived. Genesis described earth before Creation as "without form and void," with darkness "upon the face of the deep." Land divided the waters, but the waters remained. The ocean was leftover chaos, an infinite pool of the primordial stuff God had patted into order and molded into matter. It was not benign. Every raging storm, every surging tide demonstrated that the chaos at civilization's borders was trying to burst free and devour creation. It was an unspeakable horror. People feared engulfment by it. They feared the creatures that lived beneath it and the invaders that arrived upon it.

If the Biblical Flood was punishment for human sins, then it was natural to fear that the ocean was ready at any moment to again cleanse the world of humanity. People had constructed a few puny islands to protect themselves but they were constantly bombarded. Chaos threatened from every direction. Deserts were wastelands, devoid of life and hope, where a person was sure to become lost and die. Forests were dark, threatening, and dangerous lairs of monsters, ghosts, murdering bandits, and invading barbarians. No wonder Charlemagne and other rulers decreed that the forests should be cut and converted into tillable fields and pasturage.

But more terrifying than any forest or desert was the ocean. No monarch could do anything about it. Its water sickened instead of quenched, destroyed crops rather than nourished them, gave off the foul odors of corruption and death. The ocean was the absolute embodiment of chaos: an endless swirling cauldron of malignant energy that could swallow you and leave no trace you had existed.

In literature as ancient as the epics of Homer and Virgil, and through the Middle Ages into the sixteenth and seventeenth centuries, the ocean was portrayed as a wrathful horror or a spiritual abyss or the representation of hellish disorder. In storms, winds came from all four directions at once, mountainous waves laid the bottom of the sea bare in their troughs, the sky was black and rent by lightning, and the tenth wave was the most terrible one of all.

The priests of ancient Egypt so hated the sea that they would not eat fish or allow salt to be set on their tables. At the time that Spaniards were conquering the people of the Americas, Peruvians high in the Andes had a terrible fear of the Pacific, believing it caused disease. Coastal inhabitants of seventeenth-century Italy described the stinking matter washed to shore by the sea as its excrement and referred to the foam cast up by waves as its sweat. To them the tides were evidence that the sea was afflicted with fever and was breathing heavily. Elsewhere in Europe it had long been believed that the ocean was rotting and that inhaling the fetid air rising from it was unhealthy. Seasickness was a blight caused by the emanations of the spoiled sea itself, and beaches were among the most unhealthy of places, the depository of decaying waste. Even the sand on a beach was repulsive, the very opposite of the rich living soil of gardens and farmers' fields. The sea and its coast were out of harmony with the rest of creation, a clashing, muddled, uneven, and unpredictable arrangement of features too ugly to contemplate.

Those who went to sea to fish, travel, or explore knew that the spirits and deities of the water demanded tribute. The Romans would sacrifice a bull before a journey; Japanese fishermen would scatter offerings of rice; ancient Britons would toss criminals off headlands and cliffs to appease the ocean gods. It was considered bad luck to rescue a drowning person because it was assumed the sea would demand another in his place—perhaps the rescuer. A sacrifice of human blood was required at the launching of any chief's canoe in the South Seas. The Vikings dragged the keels of their longships over prisoners to soak the wood with blood. Today we sacrifice a bottle of champagne by breaking it on the bow of a ship as it is launched. It is unlucky for the ship if the bottle fails to break.

Gradually, with the exploration of the world and the discovery of the limits of the oceans, attitudes began to change. Before the seventeenth century, it was a rare European who learned to swim, and bathing in rivers, lakes, and seas was considered at best a vulgar activity for the low-bred and at worst behavior so immoral, dangerous, and bizarre it was sometimes considered evidence of lunacy. It took a health craze to alter that way of thinking. In 1621, Robert Burton's influential *The Anatomy of Melancholy* recommended swimming as an exercise to combat the affliction known throughout Europe as "spleen" (we know it today as depression or anxiety disorder), and to cure such common medical disorders of the day as "phrenzy," "nervous irritation," "vapours," "hysteria," and "nymphomania." A century later, the English author Tobias Smollett, an adamant and outspoken critic of the urban blight then tainting life in heavily polluted British cities, became fascinated with therapeutic uses of water. His 1742 treatise on hydrotherapy, *An Essay on the External Use of Water*, promoted cold baths and swimming in the waters along the French and Italian coasts. He was among the first to speak out against life in hot, filthy cities and to advocate healthy sojourns along the seashore.

Smollett and other proponents of water-therapy found precedent in ancient Roman practices, as detailed by Pliny the Elder in his *Natural History*. According to Pliny, seawater possessed many curative powers:

> It is also drunk, though not without harm to the stomach, for purging the body and for getting rid of black bile or clotted blood by vomit or stool. Some have also given it to be drunk in quartan agues, in tenesmus, and for diseased joints, keeping it for this purpose, for age takes away its injurious qualities... Sea water warmed is also injected as an enema. Nothing is preferred to it for fomenting swollen testicles, or for bad chilblains before ulceration; similarly for itching, psoriasis, and the treatment of lichens. Nits too and foul vermin on the head are treated with sea water... It is moreover known to be healing for poisonous stings, as of spiders and scorpions, and for persons wetted by the spittle of the asp *pytas*, but for these purposes it is employed hot... Swollen breasts, the viscera, and emaciation, are rectified by sea baths, deafness and headaches by the vapour of boiling sea water and vinegar.

Soon therapeutic bathing in oceans and spas was fashionable. People "taking the cure" flocked to the shore, especially along northern waters where cold water was considered more healthful than warmer waters, which it was believed had a tendency to putrefy. Seaside holidays became both healthful exercise and spiritual recreation to escape the anxieties and pressures of urban life. People suffering from ailments practiced cold bathing (the shock of the cold was said to be beneficial to ailing nervous systems) as well as drinking small amounts of seawater, which acted as a purgative to flush impurities from organs and glands.

At the end of the eighteenth century, with the advent of "natural theology," a new sensibility elevated nature and justified it as a spectacle performed by God for human benefit. Appreciating the beauty of nature and exercising the five senses became a pious act. Waves, which once were the encroachment of chaos or the punishing slaps of an angry and wrathful God, were suddenly seen as respectfully bowing to the supremacy of God and the land. The tide, instead of pulsing with threat, cleansed the shores and aided ships in navigating bays, harbors, and river mouths, and was considerate enough to recede twice a day, exposing the ocean bottom to allow people to collect the bounty of food there. For the first time, perhaps, it was possible to meet people walking for pleasure along a beach and to see houses with windows opening on a view of the ocean.

The new appreciation of the oceans reached a culmination during the Romantic Movement of the late eighteenth and early nineteenth centuries. The Romantic poets and artists glorified the ocean and the shore and frequently made it the scene for dramatic moments of self-discovery. Poems like *Childe Harold's Pilgrimage* by Lord Byron became anthems of the new appreciation of the oceans:

> There is a rapture on the lonely shore,
> There is society where none intrudes,
> By the deep sea, and music in its roar:
> I love not Man the less, but Nature more...
> Roll on, thou deep and dark blue Ocean—roll!
> And I have loved thee, Ocean! and my joy
> Of youthful sports was on thy breast to be
> Borne, like thy bubbles, onward: from a boy
> I wantoned on thy breakers—they to me
> Were a delight; and if the freshening sea
> Made them a terror-'twas a pleasing fear...

The sea satisfied the new taste for the sublime in nature, which in the Romantic era was defined as tranquility flavored with terror. Holidays on the seashore became the rage throughout Europe and a new tourism industry was born. In 1795, a French visitor to Holland's shore described shops selling "shells, stuffed fish, marine plants, artificial flowers, and above all, little models of ships, launches, and other objects related to seafaring."

Scientists, too, were discovering the oceans. During his three voyages in the late eighteenth century, James Cook charted the Pacific Ocean, sounded depths to 1,200 feet, and made detailed observations of water temperature, wind, and current. The voyages of Baron Alexander von Humboldt, Charles Darwin, and Edward Forbes added voluminously to what was known about the oceans and their flora and fauna. The science of oceanography began to develop with the work of Matthew F. Maury, a U.S. naval officer who produced the first reliable charts of ocean currents and wind patterns in 1847 and in 1855 published his monumental book, *The Physical Geography of the Sea*, with chapters on the currents, depths, winds, storms, and climate of all the oceans.

The nineteenth century's most extensive study of the oceans was begun when the British Admiralty's HMS *Challenger* set sail in December 1872 with a crew of 243, including four naturalists, a chemist, and a secretary to record their observations. By the time the naturalists returned to England in December 1876, they had traveled 78,000 miles and had visited 362 sampling stations. At each station they recorded weather observations, measured the depth, took samples of bottom sediment, collected organisms, checked water temperatures, and sampled and analyzed seawater composition on the surface and at various depths. The scientific studies were under the direction of Charles Wyville Thomson, a natural history professor at the University of Edinburgh. After his death his assistant, Sir John Murray, spent 23 years compiling the reports of the voyage. The 50 volumes and nearly 30,000 pages of observations and data that resulted provided a foundation for modern oceanography and are still used for reference and comparison.

Most of what we know about the oceans was learned in the last 100 and especially the last 50 years. We know, first, that ours is an ocean planet: Seventy-one percent of its surface is covered with water. Altogether the oceans cover more than 144 million square miles to an average depth of about 2.3 miles. The only portion of the world that contains more land than water is across a narrow strip between latitudes 45 degrees north and 70 degrees north—the northern regions of Europe, Asia, and North America. The rest is predominately ocean.

The three largest oceans of the world—the Pacific, Atlantic, and

Indian—extend north from a common source in the waters of the Southern (or Antarctic) Ocean. At the top of the globe is the Arctic Ocean. Each of these five oceans has a distinct shape and a distinct system of surface currents, and varies from the others in surface area, volume, and average depth.

The average depths of the three largest oceans are similar. The Pacific averages out as the deepest, at 14,048 feet, followed by the Indian at 13,002, and the Atlantic at 12,880 feet. The Pacific averages deepest only because it contains most of the deep ocean trenches.

Relative volumes and surface areas are not as evenly balanced. The Pacific contains 51.6 percent of all ocean water, the Atlantic 23.6 percent, and the Indian 21.2 percent. In surface area the Pacific is greatest, at 64,186,000 square miles, followed by the Atlantic with 31,862,000, and the Indian with 28,350,000 square miles. The Arctic Ocean, the smallest of the oceans, has only about one-sixth of the surface area of the Indian Ocean.

Such statistics do little to suggest the awesome magnitude of the seas. For that, only seeing them in person will do. All the oceanography in the world is nothing compared to an hour in stormy seas.

Many of the old fears and superstitions linger, and for good reason.

Storms still appear suddenly and ships still disappear without a trace. In some cultures of Africa and Madagascar it is taboo for a king to look upon the sea; a mere glimpse means certain death. On the east coast of Newfoundland, fishermen wear wool sweaters knitted with designs specific to each village, so that if a man drowns and washes up on shore his body can be more easily identified. Commercial fishermen I've talked to in Newfoundland, Nova Scotia, and Iceland admitted that they never learned to swim. There was no point, they said. When your number is up the sea will take you whether you can swim or not.

I thought of that one morning while camped on the remote and rocky southern shore of Newfoundland. It is one of the loneliest places I know, a harsh, gray coast where the spruce trees are stunted by wind and the sea is always high and the wind always cold. I stood on the beach watching waves pound the land, each deadly breaker crashing and rushing up the beach and sending stones the size of bowling balls clattering around as if they were pebbles.

A mile offshore a small fishing trawler was being tossed in the seas. It would rise on a crest, slipping sideways with the cross-waves, then plunge into a trough and disappear. The boat was not in distress—I could see crewmen working over nets in the stern—but it seemed so vulnerable and fragile that it took my breath away. I understood the fatalism of fishermen. Not even the best swimmer could last five minutes in those cold, angry waves. I shuddered and went back to camp to make a cup of coffee. Oceans are wonderful, but that morning I was very glad to be on land.

SEA LIFE

More than 80 percent of the earth's biomass, or living matter, is found in the oceans. Most of that biomass is made up of single-celled algae, which are the greatest source of oxygen in our atmosphere and directly or indirectly support most aquatic life-forms. Starting from those tiny, simple organisms, life in the sea explodes with variety and complexity.

One way to appreciate the variety of organisms in the oceans is to look at where they fit in what biologists call the trophic pyramid. The trophic pyramid is a way of illustrating the relative numbers of plants and animals at each link of a food chain. At the bottom, representing the greatest numbers and mass of organisms in the seas, are the algae and other microscopic phytoplankton known as the primary producers. Next up, and less abundant, are the primary consumers: protozoans, minute crustaceans, and other zooplankton, all of which feed on phytoplankton. Above them are carnivorous zooplankton. Next are second-level carnivorous consumers such as baitfish, which are preyed on by third-level carnivores such as salmon and cod, which finally are prey for carnivores at the top of the pyramid.

The trophic pyramid is widest at the bottom and narrows progressively as it nears the top. The greatest numbers of organisms are those at the bottom of the pyramid because it takes so many of them to support the animals higher up. On average, the transfer of energy from one level to the next is only 10 percent efficient. For a shark or killer whale at the top of the pyramid to gain 1 pound in body weight it must eat 10 pounds worth of third-level carnivore such as coho salmon. For the salmon to reach 10 pounds, it has to consume 100 pounds of baitfish. That 100 pounds of baitfish reached maturity only after eating 1,000 pounds of carnivorous zooplankton. The carnivorous zooplankton captured and ingested 10,000 pounds of herbivorous zooplankton, which in turn devoured 100,000 pounds of phytoplankton. Many thousands of pounds of organisms must be consumed to produce every pound of top carnivore swimming in the ocean. It's a messy business. A busy business. Every organism is trying to be as successful as possible at staying alive and every species is adapting accordingly. Out of all that activity came spines and venom and electric shockers, as well as a dazzling assortment of camouflages, mating systems, hunting tactics, and methods of locomotion.

The seas contain most of the living organisms on earth and representatives of every phylum, including many, such as coelenterates (jellyfish) and echinoderms (sea stars), that are not found on land. Interestingly, however, there is greater species diversity on land. Some biologists estimate that more than 90 percent of all species are terrestrial, primarily because two enormous

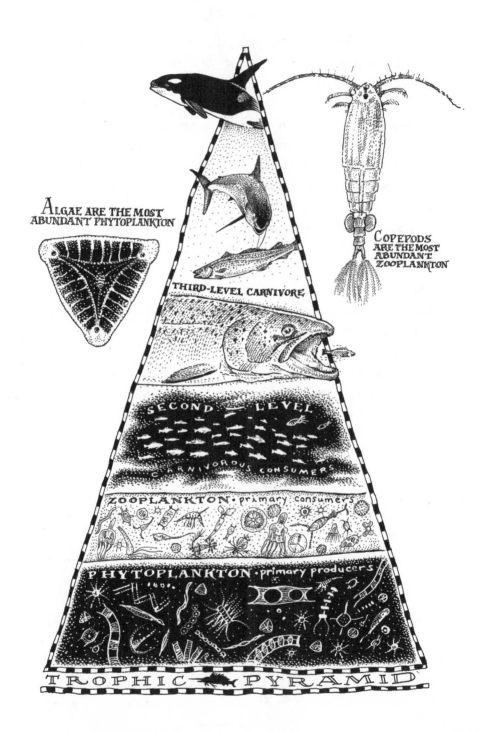

ALGAE ARE THE MOST ABUNDANT PHYTOPLANKTON

COPEPODS ARE THE MOST ABUNDANT ZOOPLANKTON

THIRD-LEVEL CARNIVORE

SECOND LEVEL CARNIVOROUS CONSUMERS

ZOOPLANKTON · primary consumers

PHYTOPLANKTON · primary producers

TROPHIC PYRAMID

groups, the insects and flowering plants, have been wildly successful on land but have few representatives in the oceans. Land habitats have sharply defined barriers, creating many isolated worlds that make it possible for, say, beetles to evolve into 25,000 species. There are barriers in the ocean as well but not as many and they are not as difficult to cross. As a result, fewer species have evolved. They are found across wide geographic ranges and often in very large numbers. The oceans are also more stable than the land, allowing species to remain unchanged for longer periods.

Marine biologists divide the oceans into several major kinds of habitats. The *intertidal zone* is the border of land and sea, and is alternately flooded and left dry by tides. Extending from the shallows to the edge of the continental shelves is the *neritic zone*. Beyond the continental shelves is the oceanic zone, which includes *surface layers* and the *deep abyssal zone*. At the bottom of the sea is the *benthic zone*.

Every zone is inhabited by organisms adapted to the special conditions there. In the most general terms, marine life is categorized into three types. Plankton are the great mass of plants and animals—such as diatoms, protozoans, and jellyfish—that either don't have the ability to move on their own or have only limited ability, and therefore float or drift with the wind and current. Among the plankton, algae are the most abundant plants, or phytoplankton. The most abundant animals, or zooplankton, are the copepods, minute crustaceans that are sometimes considered the oceans' answer to insects and form the main link between phytoplankton and the animals on the higher trophic levels. Among animals 1 to 10 millimeters long they are probably the most abundant on earth, found in densities as high as 10,000 per cubic meter in coastal waters. They're more scattered in the open ocean, perhaps 10 to 100 per cubic meter.

The nekton are the free swimmers. They include some 5,000 species of sharks, rays, tunas, squids, sea turtles, whales, dolphins, sea lions—all the fish, mammals, and reptiles that move through the oceans under their own power.

Finally are the *benthos*, the dwellers of the ocean floor. They include the seaweeds, corals, oysters, clams, barnacles, lobsters, sea worms, starfish, and all the burrowing organisms and microorganisms that dwell on the bottom of the intertidal and benthic zones.

Life in the oceans is unevenly distributed. Whales make long migrations, passing through hundreds or thousands of miles of relatively barren water to reach places where food is concentrated and abundant. The richest of those places are found in a few scattered locations near the coasts of major land masses, where upwelling currents lift a rich chowder of nutrients from the bottom to the surface. Upwelling areas are about six times more productive than the open ocean and swarm with plankton and the animals that prey on them.

One of the most productive places in all the oceans is found near the bottom of the planet, tight up against the coast of Antarctica. Every spring the cold surface water surrounding the continent is diluted with freshwater from melting ice. At the same time, warmer but denser salt water to the north slides beneath the diluted water and flows in a current toward Antarctica. When it reaches the sloping continental shelf it is driven to the surface, forming a productive upwelling known as the Antarctic Divergence. Blooms of diatoms there support enormous populations of krill and other small organisms that nourish most of the whales and other large animals of the Antarctic region.

In the fall, when the surface waters around Antarctica cool and new ice forms, another current is formed, the Antarctic Convergence. Like the Divergence, this is a current born from ice, but instead of it being diluted meltwater, its source is the brine draining from the bottom of ice as it forms. Because salinity increases the density of water, it sinks, and because it originates with ice, it is cold, nearly freezing. This very cold, very salty water plummets to the bottom and flows north across the floors of the Atlantic, Pacific, and Indian Oceans, settling into the deepest of the abysses and pushing farther and farther north until it spreads in a layer of cold and salt across the equator into the northern portions of the oceans. It retains essentially the same temperature and salinity it started with, and provides a uniform habitat for deep-water organisms across most of the deepest water of the oceans. As a consequence, many of the species of fish, corals, sponges, and isopods found miles beneath the surface near the Bahamas are found also near Antarctica.

Rivers in the Sea: Ocean Currents

Upwells and other ocean currents have a profound effect on life, not only in the oceans, but across coastal lands where climate and weather are altered by oceanic redistribution of heat. Because of ocean currents, western Europe is warmer than the same latitude in North America, and Oregon is warmer than North Korea.

Most of the great ocean currents are products of *wind-driven circulation*, and are caused by a combination of atmospheric circulation and the earth's rotation. Affecting mostly the top layers of the oceans, above the thermocline, they are relatively fast and horizontal currents. The largest and most powerful of them are gyres, large, elliptically shaped belts of current produced by the friction of prevailing westerlies, trade winds, and polar winds on the surface of the oceans. In the Southern Hemisphere, the West Wind Drift, the only circumglobal ocean current, swings continuously around Antarctica. The other major currents are constricted by landmasses.

Both hemispheres have large subtropical gyres that rotate in a roughly circular pattern, clockwise in the Northern Hemisphere and counterclockwise in the Southern Hemisphere. Smaller gyres in the low latitudes move in directions opposite of the large subtropical gyres. In the Northern Hemisphere, major gyres carry warm water north along the west sides of the oceans and carry cold water south along the east sides. The Gulf Stream, for instance, directs warm water north to Newfoundland, moderating its climate and making its offshore fishing grounds among the richest in the world. In the Pacific, cold water from the north follows the west coasts of Canada and the United States all the way to southern California.

Ocean currents are influenced by more than prevailing wind patterns, however. Both the west-to-east rotation of the planet and the Coriolis force produce circulation patterns. Centrifugal force from the earth's rotation causes water to pile up against the west side of ocean basins, concentrating

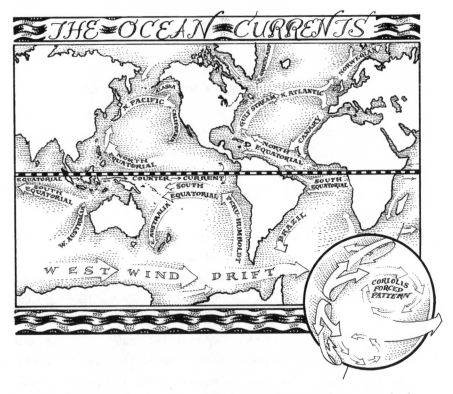

currents there. On the east side of the basins, the same force spreads the currents out, making them more diffuse. In the Northern Hemisphere, the Coriolis force—produced because of reduced global circumference at higher altitudes—causes currents that flow north along the west side of the ocean basins to be deflected east. The Gulf Stream is the best example of this. As it flows north along the East Coast of the United States it is strong and concentrated. The Coriolis force causes it to turn east near Newfoundland and cross the North Atlantic. By the time it reaches Europe it has lost much of its momentum and has dispersed into several smaller currents. In the Southern Hemisphere the effect is the same, but in opposite directions: The Atlantic's South Equatorial Current flows west to Brazil, is deflected down the coast of South America, and is turned east by the Coriolis force. It finally completes its counterclockwise rotation off the coast of Africa.

The Gulf Stream is the greatest of all the oceans' surface currents. It was noticed by Columbus and Ponce de Leon but was little understood until

Benjamin Franklin produced the first accurate chart of it in 1769, based on data he gathered during many crossings of the North Atlantic and by interviewing ships' captains. American whalers had been taking advantage of Franklin's "Gulph Stream" for years and were fond of speculating about its origins. Among the early theories was that it was an extension of the Mississippi River, which some observers calculated flowed at approximately the same velocity. Another notion, supported by Franklin, was that the trade winds caused a head of water to pile up in the Caribbean and that it escaped north under hydrostatic pressure.

The Gulf Stream has its source in the tropics, along the North Equatorial Current as it gains momentum from trade winds blowing west across the Atlantic. It flows in a powerful, steady rush from the Canary Islands to the bulging shoulder of South America, where it is deflected north. Part of the current goes northwest toward Florida, skirting the Leeward Islands and the Bahama Bank, but the major portion of it pours into the Caribbean, circles through the Gulf of Mexico, and returns to the Atlantic between Key West and Cuba as the Florida Current. It then joins with the Antilles Current to become the Gulf Stream as it flows north along the east coast of North America. It turns east across the North Atlantic, and near the coast of Europe splits into several lesser currents, the strongest of which continues southward as the Canary Current. At its maximum flow, off the coast of North Carolina, the Gulf Stream flows at 2.5 billion cubic feet per second—3,000 times the volume of the Mississippi.

Ocean currents, like river currents, create a variety of subcurrents. They can meander (especially after storms), winding slowly back and forth, traveling as much as 10 or 12 miles a day. Meanders occur frequently along the edges of the Gulf Stream where warm, north-flowing water opposes cold, south-flowing water. The meanders can become so pronounced that they break apart into isolated eddies. Such eddies typically have a diameter of 100 or 200 miles, spin at a speed of up to three feet per second, and travel a few miles a day. Off the coast of North Carolina they circulate clockwise and have warm water in their centers surrounded by spinning cold water. Elsewhere they spin counterclockwise and have a center of cold water surrounded by warm water. They often extend from the surface to a depth of one or two miles, where they produce sand ripples on the bottom, stir up blizzards of sediment, and mix the water over large areas. Most eddies lose their energy and dissipate within a few days or weeks, but a few have been followed and studied for a year or more before they dissipate.

Upwells and downwells such as those in Antarctic waters are created when wind-driven currents collide or are forced against land. A downwell, or *convergence*, results when surface water sinks. An upwell, or *divergence*, is water rising from the depths to fill the space created when currents diverge. Perennially rich fishing grounds such as those found off the west coasts of South America and Africa are a result of divergences, or upwellings, produced as trade winds consistently blow surface water away from the land.

Ocean circulation occurs also in the system of deep-water currents known as *thermohaline circulation*, which is driven by differences in water density. When water made dense by cold temperature or high salinity sinks into less dense water, it produces both surface and deep-water currents. It usually happens on a large scale: vast masses of water sliding from the surface into the deepest portions of the oceans and traveling for hundreds or thousands of miles. The Antarctic Convergence is one of the most dramatic examples. Another occurs at the Gibraltar Sill, the narrow boundary where the Mediterranean mixes with the Atlantic beyond the Straits of Gibraltar. Mediterranean water is warmer than Atlantic water but so much more saline that it is denser. When it flows across the Gibraltar Sill and enters the Atlantic it sinks to a level more than a half mile below the surface and flows in a distinct tongue of salty water for a thousand miles or more into the Atlantic.

The nineteenth-century naturalist Alexander von Humboldt was a pioneer in the study of deep-water currents. He proposed a simple pattern of ocean circulation based on the heating and cooling of water by the sun. Water near the equator was warmed, he theorized, causing it to rise to the surface and flow toward the poles, while water near the poles was cooled, sinking to the bottom and flowing toward the equator. The result should be convection cells of circulating water in both the Atlantic and Pacific

Oceans. The theory was basically valid, but far too simple. As with so many of the earth's dynamic systems, there are complications.

A more accurate model of deep-water circulation must consider such factors as salinity, bottom topography, and the force of the earth's rotation. But Humboldt was correct in one important respect: surface water in the North Atlantic, near Greenland, and in the Weddell Sea of Antarctica is made dense by cold temperatures and increased salinity and descends nearly to bottom. From there it flows in masses of cold, salty water toward the lower latitudes, is nudged by the centrifugal and Coriolis forces, and accumulates on the west side of the ocean basins.

THE ABYSS

For centuries the seas were literally unfathomable. The story goes (though it might be apocryphal) that in 1521 Magellan tried to plumb the depths of the Pacific. When he could not reach bottom after running 2,300 feet of line over the side of his ship he announced that he had found the deepest spot in the oceans. He wasn't even close.

The oceans' ultimate depths are found in the Pacific's deep-sea trenches, which are dozens of miles wide and hundreds or thousands of miles long and at least 6,000 feet deep. The deepest of them all is the Mariana Trench in the western Pacific, where its lowest point, the Challenger Deep, plummets 36,200 feet, or nearly seven miles, beneath the surface. By comparison, the highest point on land, Mount Everest, is 29,028 feet above sea level.

All deep trenches are found near volcanic islands or coast-hugging mountain ranges, where the steep land on shore continues to descend into the ocean. They are located in subduction zones where the margin of one continental plate is forced beneath the margin of another, so they tend to be in the shadow of volcanoes. The greatest distribution of trenches is found along the seaward side of the chains of volcanic islands that ring the Pacific. There the Aleutian Trench, Japan-Kuril Trench, and Mariana Trench all follow the edges of arcing island systems. The longest trench, the Peru-Chile, stretches 3,700 miles along the west coast of South America. In

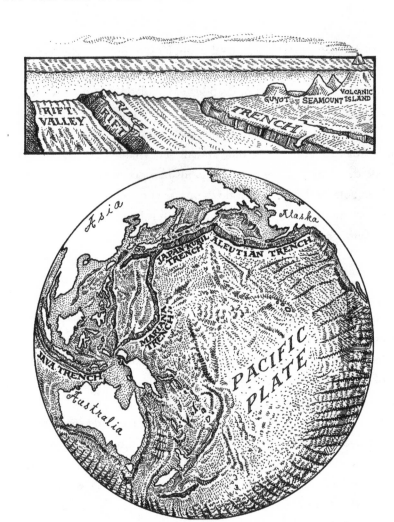

the Indian Ocean, the Java Trench runs for 2,800 miles along the coast of Indonesia. The only trenches in the Atlantic are two relatively short ones, the South Sandwich Trench and the Rico-Cayman Trench.

Even ordinary ocean depths are a largely unexplored world. The broad and relatively flat bottom of this world is known as the *abyssal floor*, and is made up of featureless plains scattered with gently rolling hills. Nowhere on the surface of the earth are there flatlands as enormous as the plains 10,000 to 20,000 feet beneath the three oceans, the Gulf of Mexico, and the Mediterranean Sea. Scattered across those plains are abyssal hills ranging from 150 to 750 feet in height. Like the foothills of terrestrial mountains, they become more numerous and higher near the oceanic ridges—those rugged spines of drowned mountains running across the oceans.

The mountains of the seas are more spectacular than any mountains of the air. The *oceanic ridges* extend in a series of ranges 50,000 miles across the ocean basins and vary from a few hundred to 7,000 feet high and up to 600 miles wide at their bases. At the crest of some of the ridges the continental plates have spread apart into central *rift valleys* 10 to 30 miles wide and up to 5,000 feet deep.

The deep-sea environment is still, cold, and dark, but much goes on down there. Deep-sea currents are sluggish but persistent. Most travel only about two to four feet in an hour, but others are swift enough to build ripples on the bottom sediment. The strongest of them flow along the west sides of ocean basins, where they are concentrated by the force exerted by the earth's rotation.

Volcanic activity is fairly common in the ocean depths, especially along the oceanic ridges and in the rift valleys, where flows of lava are common, some stretching for several miles and piling hundreds of feet above the bottom. Scattered throughout the oceans, especially the Pacific, are thousands of *seamounts*, volcanic formations similar to the cones of volcanic mountains on the surface, and *guyots*, which are seamounts with their heads cut off, probably by erosion of surface waters when the seas were at a lower level.

Life at the bottom of the sea faces a number of challenges, not the least of which is tremendous pressure. For every 10 meters of depth in the oceans, pressure increases one atmosphere. An atmosphere is the pressure on a column of mercury 30 inches high at sea level, or 14.7 pounds per square inch. If you were to dive to the bottom of the Challenger Deep, 36,000 feet below the surface, every square inch of your body would be crushed under more than eight tons of pressure.

The weight of the water at such depths would seem to make life impossible. When British naturalist Edward Forbes made the first systematic study of marine life in the 1830s, he concluded quickly that the abyss was lifeless. After dredging for specimens in the Aegean and Mediterranean and in the waters surrounding the British Isles, he named eight zones where marine life was found and declared that no organisms could exist below about 2,000 feet, which he designated the "azoic" or "no life" zone. He was convinced nothing could survive the tremendous pressure of the lightless depths. His conclusions were accepted dogmatically by much of the scientific community of his day, despite good evidence to the contrary. During his search for the Northwest Passage in 1817 and 1818, John Ross had used a "deep-sea clamm" to dredge living worms and other animals from the bottom at depths of 6,300 feet. His nephew, James Clark Ross, had similar results in even deeper water when he took bottom samples in Antarctic waters. But it was not until 1860, when a transatlantic telegraph cable was raised from the bottom for repairs and found draped with clams, corals, and other evidence of fruitful life, that biologists finally decided Forbes had been wrong.

No doubt the dogma was clung to so obstinately because it seemed to make such good sense. Life at the bottom of the sea must be able to survive incredibly difficult conditions. Besides the weight of the water, animals must deal with absolute darkness, scarcity of food, and extreme cold. There are certain advantages—cold water can hold a great deal of dissolved oxygen, and the high pressure and cold temperatures slow growth and reproductive rates, allowing organisms to live long lives—but only to the relatively few, highly specialized animals that are equipped for it. The bottom of deep-sea trenches has been found to contain about 100 organisms for every square meter, compared to more than 10,000 in the same space on the continental shelves.

SOUND UNDERWATER

The oceans are a "silent world" according to Jacques Cousteau, but he knew they were not completely silent. In fact, marine recordings reveal a tremendous range and variety of sounds, at every depth and season. In the 1950s, researchers at the Narragansett Marine Laboratory identified distinct fish calls that communicated "aggravation," "alarm," and "readiness for combat." Since then it has become clear that many fish use sounds to attract mates, stay close to one another in schools, and define and defend territories from intruders. Ocean fishes such as toadfish, drums, grunts, and croakers are named for the sounds they make by striking their hollow swim bladders with special drumming muscles along their backbones. Marine catfish bark, sea horses click, and codfish grunt. Sea bass beat their gill covers against their heads. Triggerfish, ocean sunfish, horse mackerel, and squirrelfish make rasping sounds by grinding together special teeth located in their throats. Male satinfish shiners purr to attract females. A male toadfish looking for a mate roars with a foghorn bleat loud enough to disturb the sleep of people on boats, then growls menacingly while guarding the eggs he has fertilized.

Many marine crustaceans send auditory signals to establish territories, attract mates, or warn of danger. Spiny lobsters repel intruders with a scratching sound made by rubbing the base of their antennae against their shells. A number of crab species rattle their claws when a predator comes near, causing every crab in the vicinity to duck for cover. Some of the loudest inhabitants of the oceans are only two inches long: pistol or snapping shrimps, which use their claws to produce retorts as loud as firecrackers that startle predators and stun and capture prey. In laboratories, pistol shrimp have snapped loudly enough to shatter glass jars.

Among the most celebrated of aquatic noisemakers are the whales, which sing to one another during their migrations. In *The Crystal Desert*, biologist David G. Campbell describes what he heard when he swam with humpback whales near the Galapagos Islands: "As soon as I put my head in the water, I heard their songs, seemingly coming from all directions. In the clear, warm water the songs seemed unearthly—a coloratura of trumpets, low whoops, noises like creaking hinges. The low notes were beyond the range of my hearing; I sensed them as a reverberation in my chest and lungs and cranium. These were certainly like no other animal sounds I had ever heard."

Humpback songs are as complex as birdsongs (and even sound uncannily like them when taped and speeded up 14 times faster than normal) and can last for 6 to 35 minutes, in a rhythmic pattern repeated for hours at a time. They can be heard by other humpbacks as far as 20 miles away.

Whale songs could have played a role in the myths of Sirens. Ancient sailors, hearing the eerie singing of whales through the hulls of their ships, would have been excused for giving them supernatural sources. The Sirens heard by Ulysses in the *Odyssey* could have been humpback whales, which are now extinct in the Mediterranean, though they once might have calved in the vicinity of the Sirenusian Isles in the Bay of Naples. Most whale and dolphin species communicate with a complex vocabulary of clicks, squeaks, and whistles used to attract mates and locate one another while migrating. Dolphins and other toothed whales (as opposed to the baleen or food-filtering whales) also use vocal emissions in echolocation. Like bats, they send out sound pulses that reflect back from objects, assisting in navigation and the detection of prey. When the sound waves bounce back after encountering objects, the whales gather them into sinuses in their lower jaws and process the information.

Water is a splendid medium for acoustic communication. Sound is conducted faster in dense mediums than in light ones, thus is faster in solids than liquids, and faster in liquids than gases. In the air (at a temperature of 32 degrees Fahrenheit) it travels at a velocity of about 1,089 feet per second. In seawater it travels at an average rate of 5,000 feet per second. The speed increases as salinity, temperature, and depth increase. Raising salinity one part per thousand increases the velocity of sound through it about 3 feet per second. Increasing the depth of the water increases velocity about 6 feet per second for every 300 feet of depth. Temperature affects sound even more than salinity and depth. For every degree Celsius that water is heated, the velocity of sound increases nearly 15 feet per second. Thus, sound travels best through warm, salty water at great depths.

The effect of temperature on sound velocity has special significance at the thermocline, where water of distinctly different temperatures is stratified along a distinct line. The thermocline creates a kind of channel where sounds become trapped, bouncing along like echoes in a narrow canyon. This channel has been given the name SOFAR channel, an acronym for Sound Fixing And Ranging. Sound traveling in the SOFAR channel travels more slowly than sound above and below it, but it can travel thousands of miles because there is little loss of energy within it. This interocean communication line has not been overlooked by wildlife: Fin whales broadcast their mating calls through it and are thought to be capable of hearing one another from distances of hundreds of miles.

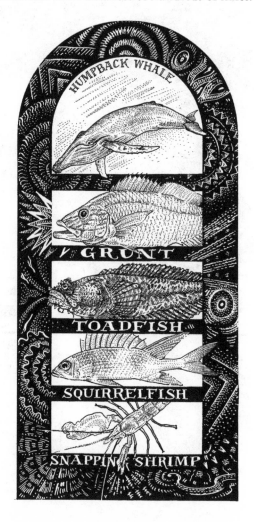

17
SALT WATER

You don't forget the taste of your first ocean. Mine was the Atlantic, at Myrtle Beach, South Carolina, during a family vacation when I was nine years old. The city of Myrtle Beach was crowded with the usual gift shops and tourist hotels but the beach along its waterfront was broad and long, big enough to match the scale of the ocean. Scattered in the sand were shells and pieces of starfish and, here and there, the black fossilized teeth of sharks. I could not believe that such treasures were lying there, ignored by everyone but me. I filled a bucket with sharks' teeth, oyster drills, quahogs, and periwinkles and searched without success for gold doubloons washed up from old pirate wrecks. Then I waded out cautiously into the low surging breakers and discovered that the water was the richest find of all. Each swell lifted me as if I were buoyed in a life jacket. A breaker washed over me and I tumbled against the coarse bottom. I came up sputtering, my eyes on fire with salt, the water filling my mouth and burning my throat. For me, a child of freshwater, this mineral water tasting of blood and copper would always be strange. What made it salty? Had it always been like this?

The defining quality of all seawater is its salinity. Every liter of water in most of the oceans and seas contains about 2.5 teaspoons of sodium chloride—common table salt. Extract all the salt from all the oceans and it would cover the lands of the earth 500 feet deep. But sodium chloride is not the only substance that makes seawater what it is. When marine chemists refer to the salt in salt water they mean all the solids dissolved in it, and there are many. More than half the elements that exist naturally on earth can be found in solution in the oceans and seas. They make their way there after first being dissolved by rivers and rainwater, or by wave erosion along the coasts, or through thermal vents at the bottoms of the oceans.

Although seawater contains many ingredients, one of its most surprising qualities is the constancy of its major components. On average it contains 3.5 percent dissolved solids, of which 99 percent is made up of six major constituents: chloride and sodium, which account for 30 parts per thousand, plus sulfate, magnesium, calcium, and potassium. Those major components come from a variety of places. Chloride and sulfate originate as gases released from volcanic eruptions and carried to the oceans by rain and rivers. Sodium, calcium, and potassium are dissolved from the mineral feldspar as it weathers in igneous rocks. Magnesium is dissolved from iron-magnesium minerals. The salinity of seawater varies, but the relative proportion of the six major ingredients remains remarkably constant from ocean to ocean and sea to sea.

More variable are the minor elements, which are defined as being present in concentrations between 1 and 200 parts per million. Most abundant of them is bicarbonate, followed in order by bromide, strontium, boron, silicon, and fluoride.

The trace elements of seawater are those found in concentrations of less than one part per million. They represent the greatest number of components, but their combined weight is only 0.01 percent of the total salts and minerals in salt water. They include lithium, aluminum, iron, zinc, arsenic, uranium, mercury, lead, nickel, copper, and silver. There is even gold in seawater, about one part per trillion, or approximately a million dollars' worth in every cubic kilometer.

The average salinity of the oceans is about 35 parts per thousand, but it varies considerably from place to place. The most obvious difference is between two types of seas: humid and arid. Such humid seas as the Pacific and Indian are flooded regularly with abundant freshwater from rain and tributaries. Arid seas, like the Persian Gulf and the Mediterranean Sea, receive relatively little freshwater, are subject to high amounts of evaporation, and therefore become unusually salty.

Hypersaline waters, with salinity over 50 parts per thousand, are found in a few isolated coastal basins where evaporation is high and there is little input of freshwater. In the open ocean salinity is highest in the arid and hot regions around 25 degrees latitude north and south. The surface water in the Sargasso Sea, for example, contains about 37 parts per thousand of salinity. The surface salinity of the mid-Mediterranean is 38 to 39 part per thousand. It reaches 42 parts per thousand in the Persian Gulf and Red Sea. In the hot but rainy tropics around 5 degrees north, salinity can be as low as 34 parts per thousand. At high latitudes, salinity varies seasonally. In the summer, melted ice tends to dilute surface water to around 32 parts per thousand; in winter, ice drains salt back into the water.

Salinity is low in the brackish waters at the mouths of rivers. Brackish water is defined as having salinity of less than 17 parts per thousand, about half that of normal seawater. The Baltic Sea is entirely brackish, with average salinity about 10 parts per thousand or less. Sailors have scooped water fresh enough to drink from the surface of the South Atlantic 50 miles off the mouth of the Amazon. Salinity varies a great deal in estuaries and in fjords, where salt and fresh waters blend. Large, drowned coastlines like Chesapeake Bay and glacially carved fjords like those common in Greenland, Norway, and New Zealand all have freshwater rivers flowing into salt water.

When freshwater is discharged into an ocean or sea strange things happen. If the estuary at a river's mouth is relatively low in energy—that is, if tides are low and there is thus little conflict with river current—the estuary becomes stratified. Freshwater is less dense, so it floats on top of salt water, which is pushed to the bottom of the estuary and advances upstream in a wedge-shaped tongue. During an ebb tide, when river current and tide combine to flow seaward, the saltwater wedge remains clearly defined.

During flood stage the conflicting currents of the incoming tide and outflowing river disturb the wedge, buckling its top into what must look something like a bunched rug, creating partial mixing of the water. Where tides are stronger, but still less powerful than the outflowing river, the estuary can be only partially stratified, as in Chesapeake Bay. In places like the Bay of Fundy, where the tide is stronger than the current of the rivers flowing into it, the estuaries tend to become totally mixed.

In some rivers and under certain conditions, salt water can move upstream for surprisingly long distances. During low-water periods on the Mississippi, for instance, a saltwater wedge on the bottom can be found as far as 60 miles upstream. But when the river is flooding and current is much more powerful, the wedge disappears at about a mile.

Significant variations in salinity usually occur only in the upper few hundred meters of the oceans and seas. Seawater below the thermocline varies little from ocean to ocean. The exceptions to that rule are a few "brine lakes" in very deep water. Very dense, salty water—up to five times saltier than the surface—has been found concentrated in bottom depressions deep in the Mediterranean, Gulf of Mexico, and Antarctic region. Perhaps the most saline of all ocean water is found 6,500 feet deep in the middle of the Red Sea. A mile-wide pool of water on the bottom there is heated by deep vents to 112 degrees Fahrenheit and contains 10 times the salinity of normal seawater.

*

Marine plants and animals rely on several important adaptations to survive in the salty environment. Their body fluids are separated from seawater by semipermeable membranes that allow some molecules to pass through while refusing others. In water, molecules tend to diffuse through a membrane until equilibrium is reached on both sides. In this process, called osmosis, molecules will always move from a region of high concentration to a region of low concentration—i.e., will move from salty water in the ocean to the less salty water in living cells, or vice versa.

The body fluids of most fish contain a salt concentration equal to about half the concentration in seawater. This mean that they lose water through their membranes as osmosis seeks to concentrate the salt in their fluids to equal the surrounding seawater. To stay in balance and prevent dehydration they must drink seawater almost constantly and secrete its salt ions across their gills, thus keeping a steady flow of half-salty water running through their bodies. Freshwater fish have the opposite problem. They have more salt in their body fluids than in the water they swim through, so must urinate copiously to rid themselves of the water being constantly diffused into their bodies.

Sharks and rays and many bottom-dwelling organisms like sponges and sea cucumbers don't bother keeping a salt balance because their body fluids contain the same amount of salt as seawater.

Many animals are limited in their geographic distribution by the salinity they can tolerate. They might be able to live in highly saline surface waters a few miles offshore, but cannot tolerate the fluctuating salinity of an estuary. The salinity in deep water, on the other hand, remains consistent

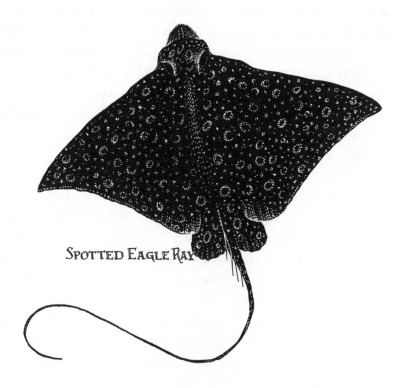

SPOTTED EAGLE RAY

throughout most of the oceans, so animals living below the surface layers are widely distributed.

Some marine animals have the ability to adapt quickly to radical differences in salinity. A salmon spends most of its life in the ocean, where it drinks seawater constantly and expels the excess salt from its gills. When sexually mature, however, it runs up a river to spawn and must undergo several metabolic changes to adapt to the freshwater. It no longer drinks water, since too much of it is already being diffused from the river through its body membranes. Instead it urinates nearly constantly to rid itself of excess water, and uses its gills—modified now—to filter additional salt ions from the freshwater of the river to supplement the salt inside its cells. After the salmon's eggs hatch, the juveniles spend months in the river waiting for the salt-secreting cells in their gills to develop. Once they're ready for salt water they move downriver to the ocean.

Dolphins and harbor seals have kidneys that can adjust quickly to variations in salinity. Whales urinate excess salt, leaving a saline trail more concentrated than their blood. Sea turtles and saltwater crocodiles (and manatees, dugong, and sea snakes) rid themselves of extra salt by weeping it from tear ducts in their eyes. Gulls, pelicans, cormorants, and other seabirds drip brine from special glands in their nasal passages.

What about us? We can't live on seawater—it overwhelms our filter systems—but if we swallow a little by mistake we quickly separate the salt from the water and urinate it or sweat it out.

*

The salinity of the oceans has stayed approximately the same during all of human history, a fact that baffled people for centuries. They wondered why the oceans did not grow less salty since rivers dumped millions of gallons of freshwater into them every day. Or they wondered why the oceans did not get more salty since only freshwater evaporates from them, leaving the salts behind.

Only in the past century or so has the fine equilibrium of the oceans been recognized. Even the very earliest oceans were salty, though probably not as salty as today. As water vapor erupted from volcanoes it carried dissolved chlorine into the sky. When the vapor condensed and fell as rain, it rained chlorine as well. Sodium meanwhile was being released from the weathering of rocks on land and finding its way through surface waters to the oceans.

The oceans have gotten more salty with age as a steady flow of dissolved minerals enters and becomes gradually more concentrated. But the change has been slow. The oceans regulate their salt balance so thriftily that the net increase is too small to measure. About four billion tons of dissolved salts enter the oceans every year via rivers, yet about the same amount is deposited in deep beds on the bottom, removed by organisms, or cast on shore by wind and waves.

The water at Myrtle Beach, then, should taste the same thirty years later. Maybe I'll sample it again someday, though I expect I'll find it less appetizing than when I was a kid. In the meantime, here in the midlands far from the nearest ocean, I mix a teaspoon of table salt in a glass of tap water and sip it. The saline content is approximately right, but the taste isn't the same.

For years the romantic idea went around that our blood contains the salty remnants of the primeval oceans where our ancestors evolved. It's nonsense, of course, but appealing nonsense. How nice to think that the moon pulls at our veins and that a swim in the ocean is a merging of inner and outer seas.

18
WAVES

I can hear waves tonight from my living room. Our house is a few hundred yards from Lake Michigan, on the east shore of a peninsula where east winds strong enough to build large breaking waves are unusual. But tonight the wind is strong and the waves are big. They come in cadence, the largest of them building to a peak and pausing, their power suspended for a moment in a silent heaving purl before they crash. They break with a hydraulic thump sudden as an ax blow, making the ground shudder, and are followed instantly by a tumbling, swirling rush of released energy that hisses like urgent low voices in a crowd. Intermediate waves run in long lapping strokes the length of the shore, like tongues licking envelopes, their breaking less sudden, announced with preambles. The sounds make me remember nights I've spent in sleeping bags on other beaches. Waking on a warm summer night to the rhythm of waves is as reassuring, you can imagine, as the sound of a maternal heartbeat to a fetus in the womb. Maybe we remember it somehow. Maybe hearing waves in darkness awakens us to our beginnings.

The waves that break against the shores of oceans and lakes were here long before our beginnings. They have been here as long as there has been liquid water and wind to loan it energy. They *are* energy, traveling through water the way sound travels through air. The energy is made audible and visible and is set in motion, a rippling of water's muscles.

How waves begin is no mystery. Children in bathtubs learn quickly that when they push water with their hands it gives way, absorbing the push and sending it rolling across the surface. Drop a pebble and the water yields to it, the surface opening a hole for a moment then springing back elastically, lifting a column of splash that rises, pauses, and plummets, sending waves outward like rings from a bull's-eye. In the complex world of wave dynamics, waves begin simply as a disturbance of the surface and a transfer of energy. And energy of one kind or another is forever disturbing water.

Watch a pond or lake and you can see waves being born. On mornings of mirror calm, the water is kept smooth by surface tension—the clinging nature of water molecules that makes it possible to float a needle on a glass of tap water. Even after the first breeze springs up the lake remains unruffled. But if the breeze continues it etches the surface with tiny waves known as ripples or capillary waves. These smallest of all water waves are also the most abundant, present not only on the ruffled surface of nearly calm water but across the slopes of large waves. They are produced when wind or raindrops or some other minor disturbing force stretches the surface film, causing it to wrinkle into waves measuring less than 1.5 centimeters from crest to crest. Like ants that can fall from 10-story buildings without harm, these half-inch waves are too small to be bothered much by gravity. Gravity can subdue larger waves to calmness, but ripples are flattened only by the cohesive force of the water molecules at the surface, the same capillary action that drives water up narrow glass tubing.

As a breeze touches a calm lake, areas of ripples appear, often in uneven, circular shapes mariners have for centuries called "cat's-paws." If the breeze continues, capillary waves offer upright surfaces that catch additional wind, like tiny sails, allowing a more efficient transfer of energy to the water. Ripples spread across the surface, forming distinct streams of disturbance or blossoming suddenly into flower shapes. Farther from shore, away from the protection of hills and trees, the wind picks up momentum and gains

THE STATE of the SEA
AN INTERNATIONAL CODE

#	WIND· KNOTS	DESCRIPTION	WAVE HEIGHT
0	<1	calm, sea like a mirror	0
1	light 1~3 air	smooth sea with ripples	0-1 ft.
2	light 4~6 breeze	small wavelets, glassy crests	1-2 ft.
3	gentle to 7~10 moderate breeze	large breaking wavelets, scattered whitecaps	2-4 ft.
4	moderate to 11~27 strong breeze	rough sea, moderate waves, whitecaps, some windblown spray	4-8 ft
5	fresh 28~40 gale	very rough sea, waves longer, heapup spray, whitecaps everywhere	8-13 ft.
6	strong 41~47 gale	high, rolling sea, foam streaks	13-20 ft
7	whole 48~55 gale	very high sea, very steep waves with overhanging crests, surface whitens	20-30 ft
8	storm 56~64	mountainous, foam-covered sea	30-45 ft.
9	hurricane >64	air filled with foam and spray, surface white, little visibility	above 45 ft.

strength, building ripples into wavelets, then into larger waves, then, if the wind is strong enough, into whitecaps. By the time the waves reach the far shore and enter shallow water they are large enough to become breakers. From the lee shore of a lake on a windy day you can watch the whole process: calm water before you, ripples a short distance out, larger waves rushing away across the open lake.

Once acted upon by a generating force such as wind or a passing boat, large waves drive forward and gravity tries to fill every trough with a crest. This chasing of crest after trough is an effort to achieve the equilibrium of a flat surface. Gravity wants to create calm seas, which is why most waves larger than ripples are known as *gravity waves*.

Most gravity waves are produced by wind, and their size is determined by the wind's speed, duration, and "fetch," or the distance it blows unhindered. An increase in any of those factors results in increased *wave height*, which is the distance from the bottom of a trough to the top of a crest; greater *wave length*, which is the distance from one crest to the next; and longer *period*, which is the time it takes a wave to pass from crest to crest.

Wind can produce several distinctive kinds of waves. Ripples grow into wavelets, then into increasingly larger waves. In the open ocean, *sea waves* (also called chop) are always under the direct influence of the wind and are likely to have smooth, rounded troughs between sharply peaked crests. They often occur in complex patterns, with waves of many sizes coming from several directions, creating confused, choppy, unpredictable seas. The roughest seas are typically composed of short, steep sea waves produced by storms.

Swell waves travel beyond the wind, either leaving the area where it is blowing or continuing on after the wind has ceased. They are usually low and long—with low wave heights and long wave lengths—and follow each other at approximately the same distance. Such waves begin when sea waves "decay," their crests becoming lower and more rounded and symmetrical until they proceed as an orderly series of similarly sized waves. In outline, swells are shaped much like a true sine curve, the classic wave of laboratory theory. It's an efficient shape: As long as they remain in deep water, swells can cross thousands of miles of ocean and lose very little energy. They are among the longest of all waves, not uncommonly measuring 1,000 feet from crest to crest and, on rare occasions, measuring as much as a mile. They are also faster than other waves. In the deep water of the open ocean they travel at 15 to 20 miles per hour, which allows them to pass through slower sea waves and organize themselves into clusters called *trains*. Such clusters eventually crash to shore as a succession of large breakers, perhaps contributing to the myth of the "ninth wave." Many people have believed that the ninth wave (or tenth or third, depending on the place) is always the largest. An hour of observation on any beach should be enough to disprove the myth, though it lives on. Some waves are larger than others, true, but the disparity is caused because swells and sea waves sometimes pass into one another, combining their energy and producing random waves much larger than average.

The steepness of a wave never exceeds a 1:7 ratio of height to length. Thus a wave measuring seven feet long from crest to crest cannot be more than one foot high. *Whitecaps* are formed when wind stronger than about 20 feet per second (13.6 miles per hour) pushes small steep waves to that 1:7 limit. When the limit is reached the waves become unstable and break into a froth of turbulence.

In a whitecap or breaking wave, particles of water fall down the crest and make actual progress. But in ordinary waves, water particles move only in well-defined circular orbits, essentially remaining in place while the wave passes through them. This motion is one of the oddest and least understood features of a wave. Waves are energy passing through water, visible only on the surface, but they have roots extending downward a distance equal to one-half the wave's length. As the energy passes it sets particles of water spinning in circular orbits, like rollers on a conveyor that spin when a crate passes over them but remain seated in place. The orbit of the water particles is large at the surface, where the diameter is equal to the height of the wave, but diminishes quickly with depth. Beneath the final, smallest orbit at the bottom of the wave there is no motion at all.

This difficult concept can be tested by filling a plastic bottle with water so that it barely floats. Place the bottle in waves next to a stationary object and note how it moves slightly forward with every crest and slightly backward with every trough. The orbiting water particles within each wave follow the same swaying motion.

When waves enter shallow water they undergo dramatic changes. At a depth equal to half of each wave's length, the lowest of the orbiting particles come in contact with the bottom and, instead of orbiting, begin to move back and forth. The particles above them continue to revolve, but their path is elliptical rather than circular, a restriction of free movement that causes an accordion-style compression of the long and rounded crest into a short and steep-sided peak. Steep waves are unstable. When the crest travels faster than the lagging bottom it tumbles forward and collapses.

Waves breaking onshore can take three forms. A *spilling breaker* collapses slowly over a long distance, with its crest spilling forward down the front of the wave as it progresses. A *plunging breaker* gradually steepens as it enters shallow water, then breaks in a sudden, instantaneous crash as its top plunges forward. A *surging breaker* rushes or surges up a beach without first growing steep.

The types of breakers are determined by a number of factors. Plunging breakers are only possible under ideal conditions. Bottom must be smooth and uniform; currents and wind must not create instability in the wave. Most of the time, on most beaches, conflicting forces slow the wave and cause it to break gradually in a spilling or surging breaker. But if allowed to break all at once, the effect can be spectacular. All the energy the wave has carried for hundreds or thousands of miles is poised to explode in a single, shattering moment. Water and air meet land in a burst of spray, with noise that can be heard a mile or more away.

You can get a fairly accurate picture of the bottom contours of a coast by watching wave behavior. Large waves that do not break until they are very near shore indicate deep water; breakers far out mean shallow water. Waves that break offshore then re-form and continue on before breaking again probably indicate a submerged sandbar or other structure. Above the sandbar the waves become steeper and break, but as they enter the deeper water in the trough between the bar and the shore (or between a succession of bars) they lengthen and become stable enough to cease breaking.

Like waves of light, water waves can be diffracted, reflected, and refracted when they encounter stationary objects or shoreline features. In *diffraction*, waves bend around a barrier such as a jetty or sandbar, producing smaller waves behind it. It explains why small islands in mid-ocean have no protected shore. Heavy surf will pound even their lee sides, making it difficult to land boats.

Waves *reflect* when they meet upright jetties, seawalls, or other vertical walls. Reflected waves rebound back into the path of oncoming waves, sometimes reinforcing an advancing wave and making it larger. They might also meet oncoming waves and halt them, turning them into *standing waves* that move up and down but make no horizontal progress. If waves meet a barrier at an angle they are reflected away at the same angle. Whenever a wave meets a beach most of its energy is absorbed, but there is always some reflection.

Refraction occurs when waves are bent by shallow water. Waves rarely strike a beach straight on because one part of a wave always comes up against shallow water before other parts. The first part to touch bottom slows, forcing the wave to turn at an angle to shore, causing breakers to proceed from one end of the beach all the way down to the other rather than breaking everywhere at the same moment. When waves follow one after another at a consistent angle of refraction, they produce consistent *longshore currents* running parallel to the front of the beach. These currents carry a great deal of sediment along the shoreline and are an important force in building sandbars and other features of a beach.

Driftwood left stranded high on a beach has been deposited there by a somewhat different force than gravity waves or current. Objects floating in waves follow the oscillating pattern of the water particles, traveling slightly forward with each crest, returning slightly backward with each trough. They make progress, however, because they are subject to the power of currents beneath the surface and winds above it. Once a piece of driftwood reaches the area of breaking waves at shore it is propelled shoreward by *waves of translation*, or bores. These result when ordinary waves break in shallow water, dumping a sudden mass of water onto the shallow surface between breakers and sending a low, flat bore racing forward up the beach. Such shoreward-running waves have a crest but no trough, and are composed of water that actually passes forward as the wave progresses, making a "translation" of water from the breaker to the beach. Objects floating in waves near shore get caught in this translation and are pushed up high and dry into beachcomber country.

The writer Henry Beston, who spent a year living on the shore of Cape Cod before writing his nature classic, *The Outermost House*, proposed that the best time to be on a beach was when the surf was high. It was then, he wrote, that the breaking waves seemed to charge shore like warhorses: "As they approach, the wind meets them in a shock of war, the chargers rear but go on, and the wind blows back their manes. North and south, I watch them coursing in, the manes of white, sun-brilliant spray streaming behind them for thirty and even forty feet. Sea horses do men call such waves on every coast of the world."

Tonight on my own small coast of the world the waves charging to the shore are steady and insistent and fill the night with the noise of culminating forces. I listen to the thump of breaking waves and realize that each wave borrows its sound from the wind, which has plenty to spare, then alters it to match its shape. It is a sound with crests and troughs in it, with rhythms as elemental as the beating of our hearts and as ancient as the phases of the moon. Wind has always paired with water to make such sound. Waves have always ended this way.

FREAK WAVES AND ROGUES

Waves generate much sound and fury, impressing us greatly, so we can probably be forgiven for exaggerating their size. I once spent a miserable day tossed around in a small cod boat off the southern coast of Iceland by waves I was certain were at least 12 feet high. They came at us like they were attacking, rising out of the choppy chaos of smaller waves until they

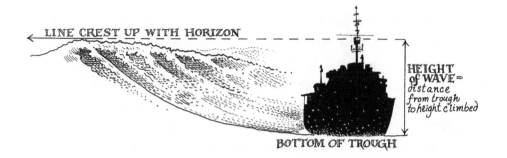

LINE CREST UP WITH HORIZON

HEIGHT
of WAVE ≈
distance
from trough
to height climbed)

BOTTOM OF TROUGH

crested above the height of the cabin roof and seemed about to come
down like giant fists. It was humbling and a little disappointing to learn
later from the weather service that the waves had been mere six-footers.
I've taken some comfort since watching anglers on Lake Michigan motor
into harbors for protection from waves measuring barely three feet from
trough to crest.

Even in the open ocean the immense waves of popular imagination
are rarely encountered. At least three-quarters of all sea waves have a height
less than 12 feet, and only very powerful storms produce mountainous seas
of more than 30 feet. Nonetheless, some parts of the oceans have earned
notoriety from producing larger-than-average waves, some of them very
large indeed. Many of the biggest waves that reach the shores of Europe
and eastern North America originate in the middle of the North Atlantic,
especially in the stormy regions south of Iceland and Greenland. The long
rollers that frequently strike the southwest corner of England can begin as
far south as the Cape of Good Hope, South Africa—some 6,000 miles away.
In the Pacific, swells washing on the shores of California and even Alaska
have been traced to the Antarctic Ocean. The winds south of Cape Horn,
at the southern tip of South America, follow the "Screaming Sixties" all
the way around the world without interruption by continents. This endless
fetch produces long, powerful waves known as "graybeards" or "Cape Horn
Rollers" by mariners who claim they have seen them measure a mile from
crest to crest and reach heights of 200 feet.

Such figures were once discounted as the result of fancy and fear by
oceanographers convinced that no wave could surpass 60 feet in height

without collapsing. But this "60-foot rule" has been challenged recently by computer models that seem to prove that powerful storms are capable of creating waves as high as 219 feet.

The big waves of computer models can be an exercise in theory divorced from practice. In nature, waves are not nearly as predictable and orderly as they are in laboratories, computers, and mathematical equations. While features such as velocity, period, and length are easy to measure in controlled circumstances, in the open ocean they can be hopelessly confused. Swells the size of small hills come at you from several directions at once, each rising and descending unpredictably, the slopes hacked by wind waves and covered with a perplexity of tiny capillary waves. One moment a breaking curler rushes down the slope of a large swell, then a trough fills with the crest of a wave coming from the side or falls away into yet a deeper trough.

The random and disorderly nature of waves in the oceans almost guarantees that they will occasionally produce waves of a higher order. Such "rogue," "freak," or "episodic" waves are known as Extreme Storm Waves (or ESWs) when they develop within a storm, and often rise in groups of three that have earned the quasi-mythical designation "Three Sisters Waves." Most reports of ESWs note that they look like "walls of water," that they're accompanied by steep-sided and unusually deep troughs (deep enough to occasionally swallow a ship), and that they spill forward and break simultaneously and with equal energy the entire lengths of their crests. An enormous freak wave with those characteristics struck the *Queen Mary* amidships south of Newfoundland while ferrying U.S. soldiers home at the end of World War II, rolling her to within a degree or two of capsizing.

Rogue waves can occur on relatively calm seas with no storms for hundreds of miles. Trains of swells traveling in the same direction but at different speeds will pass through one another; when their crests, troughs, and lengths happen to coincide they reinforce each other, combining their energies to form unusually large waves that tower mountainously for a few minutes then subside. Such giants can suddenly reach several times the height of most of the waves around them, forming mid-ocean breakers that are probably responsible for at least some mysterious disappearances of ships. When such anomalous waves occur near shore, roaring down without warning on piers and rocky ledges, they sometimes sweep people to their deaths.

As waves meet opposition from an oncoming current the wavelength is shortened and the wave is forced to become steeper and as much as two to four times higher. Examples can be seen on a relatively small scale at the mouths of rivers or estuaries, where the outflowing current creates high choppy waves that can be dangerous for small craft. When a current exceeding about four miles per hour is spread over a large area and involves massive amounts of water the danger is magnified. Regions notorious for such hazardous juxtaposition of waves and current include the Agulhas Current off the east coast of South Africa, the Kuroshio Current southeast of Japan, and the Straits of Florida and elsewhere along the Gulf Stream, where 30-foot waves blown up by nor'easters sometimes reach 40 to 60 feet in height.

Reports of rogue waves 100 or more feet in height are no doubt often exaggerated, but careful measurements by coolheaded observers have left little doubt that waves of that size sometimes occur. One fairly reliable way sailors have measured such waves is to climb the rigging until they reach the point, while the ship is in the bottom of a trough, at which the crest of the wave lines up with the horizon. The height climbed then equals the height of the wave. This technique, or variations on it, have been used to measure some waves of terrifying dimensions.

One of the largest waves ever measured was reported by Navy Lieutenant Commander R. P. Whitemarsh in a famous paper titled "Great Sea Waves," published in the *U.S. Naval Institute Proceedings* in August 1934. Whitemarsh encountered a storm on February 7, 1933, while crossing the Pacific in the navy tanker the U.S.S. *Ramapo*. During this storm, west winds had a fetch across thousands of miles of unobstructed ocean and had already blown at gale force for days. With the wind at 60 to 66 knots (69 to 76 miles per hour) directly from the stern and the waves consistent, without cross seas, Whitemarsh and his crew were able to estimate (based on the 478-foot length of the ship) that the length of the waves was 1,000 to 1,500 feet. Using a stopwatch they timed wave periods up to 14.8 seconds. Crew members standing on the ship's bridge could measure the height of a wave by lining up its crest with the horizon and a point on the ship's mast (making the line of sight approximately horizontal) while the stern of the ship was at the bottom of a trough. By triangulating from the line of sight to the bottom of the ship's stern (or the trough of the wave) they were able to determine the height of each wave. The largest they measured was 112 feet.

Other notable waves have been measured against stationary objects, especially lighthouses. In the winter of 1861, for example, the Bishop Lighthouse in the Isles of Scilly was struck by a wave high enough to smash a fog bell hanging 100 feet above the ground (eyewitnesses insisted the crest of the wave came *down* on the bell), breaking a metal support bracket four inches thick and tossing the bell to the rocks below where it smashed to pieces. Waves have occasionally passed over the top of the 75-foot-high lighthouse on Minots Ledge, Massachusetts, and have flung rocks through the glass of Oregon's Tillamook Rock Lighthouse, 133 feet above the water.

OIL ON THE WATER

According to the folklore of the sea, oil calms troubled water. In popular use the expression refers to the way that tact or gentle reasoning can soothe tempers and restore calm after an argument. But the expression is also literally true.

In traditional seamanship, the best way to calm a choppy sea was to fill a canvas bag with rags that had been soaked in cod-liver oil or whale oil and hang it over the side of the boat. It was important to use fish and animal oils—petroleum did little or no good—and to allow it to spread slowly, drop by drop across the surface. Even an amount as small as half a gallon an hour was said to calm the water for 100 feet or so around the boat. Small waves were eliminated altogether and the crests of larger waves were rounded and prevented from breaking.

Although no ancient mariner could have known it, waves are flattened by a thin layer of oil because it forms a membrane that remains strong and flexible even when only a millionth of a millimeter thick. The membrane, though just a few molecules in depth, adds to the natural surface tension of the water and dampens the energy contained in waves. The secret is to keep it thin: A thick layer of oil actually has less surface tension than a thin layer.

For centuries sailors have noticed that choppy waves can also be subdued, though for a different reason, by a good hard thrashing with rain. The impact of a single raindrop striking the surface of a body of water creates a splash followed immediately by a tiny eddy beneath it. When the entire surface is covered with such splashes and eddies their combined turbulence stirs the water beneath the surface and interrupts the organized, orbital motion of the waves. The waves then die a natural death and the sea flattens.

STORM SURGE

When coastal regions are struck by great ocean storms, most of the destruction is caused not by wind but water pushed ahead of the storm like snow before a shovel. When this shoved water, or *storm surge*, reaches shore it is often called a "storm tide" or "tidal wave," though it has no more connection with gravitational tide than does a tsunami.

The strong winds of tropical cyclones—known variously around the world as hurricanes, typhoons, or cyclones—create waves driven so relentlessly forward that they begin to crowd upon one another, piling up when they reach shallow water and creating a continuous current rushing landward faster than it can return to the sea. Hurricanes with winds of 75 to 95 miles per hour will typically produce a storm surge about 5 feet high, while winds of 130 to 155 miles per hour can be preceded by a storm surge as high as 18 feet. All that water surging over the land becomes a foundation for waves, which in turn cause more flooding. Such flooding can extend miles inland.

Storm surges are enhanced by the extremely low atmospheric pressure that accompanies tropical cyclones. The low pressure at the center of the

storm causes the water beneath it to rise in a "hump" that floods inland as the storm strikes a coast.

The Galveston, Texas, hurricane of 1900 is an example of how destructive the combination of storm surge and large waves can be. That storm brought winds of 120 miles per hour and storm surge 15 feet higher than the usual high tide. The surge carried 25-foot waves into Galveston, leveling much of the city and drowning 5,000 people. Even more terrible was the 30-foot storm surge that swept inland along the Bay of Bengal in November 1970, killing an estimated 200,000 to 300,000 people in Bangladesh.

Waves Underwater

Dolphins riding effortlessly beneath the bow of a moving ship are applying the same principles used by humans who surf the breakers off Honolulu, with an important difference: The wave they are catching is an underwater one. The pressure of water being pushed ahead of a ship creates a constant underwater wave shaped much like a surface wave. Dolphins learn to tilt themselves at a slope that matches the slope of those submarine waves, so that, in effect, they're on a perpetual downhill ride. Slight adjustments of body angle allow them to surface periodically for air without losing velocity, then to return to their position on the front slope of the wave.

So-called "internal" or "submarine" waves also occur deep underwater when masses of water of different temperature or density strike forcefully against one another. They can be generated by earthquakes and underwater volcanic eruptions, or can result from tides or other deep currents. They've been measured over 300 feet high, with wavelengths of more than 1,000 feet, and with periods lasting several minutes. You can generate them yourself with oil and vinegar combined in a bottle. Tip the bottle and small internal waves will travel lazily back and forth across the boundary where the oil and vinegar meet.

When a slow-moving ship enters the mouth of a large river, estuary, or fjord, it sometimes encounters mysterious resistance that can stop it dead in the water. Sailing vessels sometimes become unmanageable, refusing to respond to the tiller, and motorized vessels sometimes lose speed so abruptly they stop making progress and are occasionally stranded.

Mariners of old commonly believed that the phenomenon of "dead water" was caused by a "crust" of freshwater sticking to the vessel, slowing it the way it would be slowed if the entire hull was coated with a thick layer of barnacles. Ancient accounts blamed remora, those sucker-lipped fish so often seen attached to the sides of sharks and other large fish. For centuries, attempts to break free of dead water included having the entire crew run repeatedly forward and aft along the deck, firing guns into the water, scooping quantities of seawater over the deck, pouring oil on the water ahead of the vessel, dragging a hawser from bow to stern beneath the ship's hull, working the rudder rapidly back and forth, and slashing and beating the water beside the ship with oars and other tools.

When tugboats with a vessel in tow encountered dead water they discovered that several tactics worked. The simplest was to stop for a few minutes, allowing the stern waves to pass, then proceed at full speed. Another was to shorten the towrope as much as possible to allow the tug's screw to mix the water around the towed vessel.

The phenomenon of sticky or dead water can be explained by

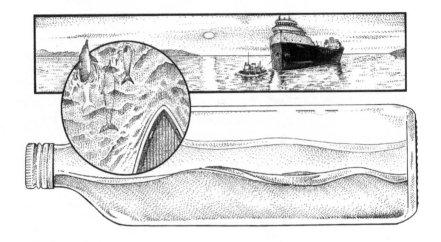

underwater waves. Dead water occurs only where there are layers of water of very different density, as when fresh or brackish water rests on top of salt water, or where warm water is on top of cold water. When portions of a vessel's hull are traveling in water of different densities, the disturbance creates submarine waves that bound off the interface of the two densities of water and create a zone of turbulence that increases resistance. In some places, such as the Norwegian fjords, where freshwater on the surface rests above very dense saltwater a few feet below, the dead-water phenomenon is particularly noticeable. If the layers of water are mixed by wind, however, the effect is lessened and will not reoccur until the water has been calm for several days, allowing fresh and salt waters to stratify. Sailors have learned that accelerating their vessels before they reach dead water until they exceed about four miles per hour, the speed of submarine waves, allows them to break free of even the stickiest water.

Tsunamis

The most powerful and destructive of all ocean waves are the seismic waves known popularly as "tidal" waves. These most massive of all waves have nothing to do with tides and everything to do with violent disturbances in the oceans. Oceanographers are nearly unanimous in calling them by the Japanese word, *tsunami,* which translates as "harbor wave," in reference to their tremendous impact when they enter harbors, ports, bays, and other confined shoreline areas. They've devastated coastal regions on every continent and have earned a possible role in many of the flood myths found in virtually every culture. Historians have suggested them as the cause of events as disparate as the destruction of the Minoan civilization on Crete, the disappearance of the legendary city of Atlantis, and the parting of the Red Sea by Moses.

Most tsunamis are formed when submarine earthquakes registering 6.5 or higher on the Richter scale create a sudden displacement of water. At the epicenter of the quake on the ocean floor there is likely to be abrupt vertical movement of the earth's crust as one side of a fault is suddenly thrust upward. That upward movement pushes a dome of water to the surface, where it sends waves radiating outward. Or a mass of rock along a fault line on the bottom can suddenly drop, causing water in a column all the way to the surface to drop also, creating a hole on the surface. When surrounding water rushes in to fill the hole it sets off a series of up-and-down oscillations that send waves radiating outward. The effect is rather similar, on a grand scale, to dropping a pebble into a still pool. Even "slow-slip" quakes that might shift oceanic plates only three feet in a couple of minutes can produce destructive tsunamis. Called "tsunami earthquakes" by seismologists, they were too subtle be detected on seismic equipment until the 1980s, yet they release vast amounts of energy into long-period waves. On September 1, 1992, a quake of this kind occurred undetected along a 60-mile fault in a deep trench off the coast of Nicaragua. It sent a 30-foot-high tsunami crashing into a 200-mile stretch of Nicaraguan shoreline, destroying 13,000 houses and killing 170 people.

Although earthquakes are the most common cause of seismic waves, they can be created by other forces, including nuclear explosions, landslides (both above and below the surface), the impact of large meteorites, and volcanoes. When the volcano at the center of the island of Krakatau erupted in the Dutch East Indies on August 27, 1883, it blew 10 square miles of land into the air. Instead of an island with an average elevation of 700 feet, there was suddenly an enormous steaming hole 900 feet below sea level. Water rushed in and waves rushed out, radiating from the explosion in a series of seismic waves that swept across the Indian Ocean, rounded the Cape of Good Hope, and traveled north the entire length of the Atlantic. They were measurable as far as Panama, 11,470 miles away. Closer to the eruption, tsunamis at least 70 feet high (and perhaps as high as 140 feet) struck the shores of Java and Sumatra, annihilating hundreds of towns and villages. At one seaside village, where all but one of the 2,700 residents were drowned, waves erased all trace of the village then swept away several stone houses at the top of a 115-foot-high hill beyond it. A ship's engineer witnessed the destruction of another coastal town moments after a tsunami passed beneath his ship, the *Loudon*: "Before our eyes this terrifying upheaval of the sea, in a sweeping transit, consumed in one instant the ruin of the town; the lighthouse fell in one piece, and all the houses of the town were swept away in one blow like a castle of cards. All was finished. There, where a few moments ago lived the town of Telok Betong, was nothing but the open sea." In all, more than 36,000 people drowned.

A tsunami receives its energy from powerful geological forces, but rather than manifest that energy in a high wave surging across the ocean, it adopts a more efficient form. The wave can cross thousands of miles of ocean at speeds of approximately 500 miles per hour, but it does it in a crest only a few inches to a few feet in height, extended across a wavelength that can reach more than 150 miles. Instead of traveling just on the surface, like a wind wave, a tsunami extends all the way to the bottom. The resulting wave, although powerful and fast, is so long and low that it can pass beneath the hull of a ship in mid-ocean without anyone on board noticing. Before scientists began using sensitive seismic devices to monitor them, tsunamis could cross entire oceans without being detected.

But as a tsunami nears shore it becomes terribly evident. As the waves enters shallow water it is slowed by friction along the bottom, forcing the wavelength to shorten and energy to be compressed into height instead of length. The wave slows from the 500 miles per hour of the open ocean to between 30 and 100 miles per hour. As it slows it rises

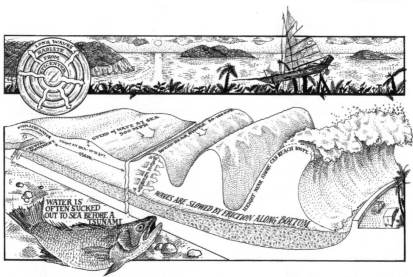

until the wave that was silent and scarcely noticeable becomes a monster, a virtual mountain of water. When it strikes—especially after funneling into a bay or harbor with a long, gradually sloped bottom—the results can be devastating.

The first sign of an approaching tsunami is sometimes a gradual, steady recession of the shoreline as the trough of the wave arrives. Bays and harbors might become emptied of all their water, leaving boats stranded on bottom and fish flopping helplessly. Such an unusual sight has cost many curious beachcombers their lives. Unable to resist the novelty of walking on the bottom of the sea they wander out from shore. But 10 to 30 minutes later the water returns, with a vengeance.

A ship is at risk from a tsunami only if it is caught in port. The best defense is to pull anchor and put to sea. Observers on vessels a mile or two offshore have reported watching powerful tsunamis breaking onshore without having noticed any unusual waves pass beneath them.

Among the most destructive incidents in the history of our planet must certainly have been the tsunamis produced when a gigantic asteroid struck the ancient Caribbean Sea about 65 million years ago. The remnants of that impact, including a 100-mile-wide crater buried beneath the Yucatan Peninsula, suggest that the waves thrown outward by the impact were hundreds, even thousands, of feet high, and inundated most of the continental coastlines, contributing perhaps to the widespread extinctions of dinosaurs and other animals that occurred at the boundary of the Cretaceous and Tertiary periods. Some geologists have suggested that another asteroid struck the Atlantic about 40 million years ago and sent thousand-foot waves crashing across what is now Maryland and Virginia.

The earliest reported tsunami destroyed Amnisos, Greece, about 1470 B.C. Among the most destructive in recorded history were a series of great waves that inundated Constantinople, Turkey, following an earthquake on September 14, 1509; an earthquake near Lisbon, Portugal, on November 1, 1755, which sent 15- to 40-foot waves pounding the coasts of Portugal and Spain, drowning thousands; and a submarine earthquake on the Grand Banks in the North Atlantic on November 18, 1929, which sent a tsunami 50 feet high roaring the length of narrow bays along the Burin Peninsula in Newfoundland, destroying villages and taking many lives.

On August 13,1868, in Iquique, Peru (now northern Chile), a wave with a reported height of 70 feet swept without warning across the city. When it receded it emptied the Bay of Iquique, exposing a bottom normally covered with 24 feet of water. Within minutes a 40-foot wave followed. A side-wheeler gunboat, the USS *Wateree*, anchored in the harbor at nearby Arica when the waves struck, was carried completely beyond the town, scraping the tops of buildings with its hull. It was left stranded a mile inland.

Japan is battered by more tsunamis than any other place (followed by Chile and Hawaii) and has a long and brutal history of devastation by the giant waves. Located along the Pacific Rim, near the most active seismic and volcanic regions in the world, the islands of Japan with their many natural harbors and inlets are particularly vulnerable. More than 100,000 Japanese were drowned by a tsunami that struck in 1702. Another, in 1771, lifted coral heads from offshore reefs and carried them— some weighing an estimated 750 tons each—several miles inland, where they remain to this day; the highest wave that day was said to be 260 feet high and killed some 11,000 people. On June 25, 1896, an earthquake rattled the bottom of the ocean in the Tuscarora Deep, 700 miles from Japan; an hour later a tsunami 100 to 110 feet high took more than

27,000 lives and destroyed 13,000 homes along a 200-mile length of Japan's northeastern coast.

After a train of tsunamis struck the Hawaiian Islands and killed more than 160 people on April 1, 1946, a group of oceanographers organized and formed the Seismic Sea Wave Warning System (SSWWS). This was the first of several warning systems that rely on seismic equipment and tidal monitors to detect and pinpoint the locations of earthquakes powerful enough to produce tsunamis and quickly issue "tsunami watches" and "tsunami warnings" to areas that might be in their path. Since then early detection of tsunamis and quick evacuation of threatened regions have reduced the deadly impact of the waves.

But some tsunamis are so enormous and arrive so suddenly that not even early warnings can prevent catastrophe. About 18,000 people were killed on March 22, 2011 when a magnitude 9.0 earthquake 80 miles off the coast of the Japanese city of Sendai, triggering a tsunami that washed over the east coast of Japan. But that was miniscule compared to the massive tsunamis that rushed across the Indian Ocean and inundated islands and coastal regions of Indonesia on December 26, 2004. The U.S. Geological Survey later estimated that the 9.1 magnitude earthquake that caused the Indian Ocean tsunamis released an amount of energy equal to 23,000 Hiroshima-sized atomic bombs. An estimated 260,000 died .

Tsunamis are among the most awesome and terrifying of natural phenomena. Little wonder, then, that the Tlinget Indians along Alaska's coast believe that when spirits in the earth are angry with humans, they send giant waves to punish them.

19

THE TUG OF THE MOON:
OCEAN TIDES

In that corner of the Bay of Fundy known as the Minas Basin, my wife and sons and I explored muddy creeks and small isolated pools scattered across a wasteland of mud. We knew that somewhere to the west was open water, but we could see nothing but mudflats all the way to the horizon. Nova Scotia seemed as remote from the ocean as Nevada. That was in the morning.

In the afternoon, when we returned, the land of mud and puddles was gone. In its place was the Atlantic.

The Bay of Fundy is justifiably famous for having some of the highest tides in the world. Until you see the ocean crawl toward you across miles of tidal flats and finally break as surf at your feet, you can hardly imagine what the fuss is about. Or else you over imagine it and expect the tide to thunder up the narrow bay with the ferocity of a tsunami. Tides are rarely that dramatic. In Fundy they make slow and incremental progress, like the rising moon, coming across the mudflats a little faster than you can comfortably walk. But at their peak they can add 40 feet of water to the bay.

The tide is the ocean's longest wave—a wave so large that it travels in a massive, rhythmic slosh around the world, twice a day, regular as clockwork. This wave has a period of 12 hours and 25 minutes and a length half the circumference of the planet. When the crest reaches shore it is called high tide; when the trough arrives it is called low tide. The rising crest *floods* the shore; the falling crest *ebbs* away from it.

The power to run the tide comes from a combination of centrifugal force and the gravitational attractions of the moon, earth, and sun. Mass and distance are the determinants of any gravitational equation. The sun has enormous mass, but it is so distant from earth that the much smaller moon exerts twice the gravitational draw on the oceans. Thus the tide rises and falls to a lunar rhythm.

Its schedule is lunar as well. The earth rotates on its axis in the same direction that the moon travels during its 29.5-day orbit of the earth. For a point on earth to rotate once beneath the moon it must complete a 24-hour rotation plus a little more—about 50 minutes—to make up for the distance the moon traveled in the same amount of time. The tidal day, then, is 24 hours and 50 minutes long, making each day's two high tides 12 hours and 25 minutes apart. If a tide peaked yesterday at 7:00, it will peak today at 7:50.

Tide waves are bulges of water protruding from the sides of the earth. One bulge appears directly beneath the moon, pulled by its gravitational attraction. At the same time there is an equally large bulge on the opposite side of the earth. This second bulge is not a product of gravity, but of centrifugal force. If a child takes your hands and spins in circles around you, centrifugal force shifts your center of balance toward the child. Though you are much heavier than the child, you must lean back to compensate for the force pulling on your arms. The two of you then spin on an axis somewhere between your ordinary center of balance and the child's hands.

In a similar way, as the moon spins around the earth it pulls the earth's center of mass closer to the moon. The earth still rotates on its central axis, but its mass in relation to the moon is off center. The centrifugal force created as the moon and earth lock hands and swing in their monthly dance causes the ocean to bulge on the side of the earth opposite the moon. So

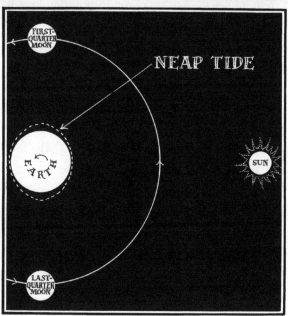

there are two equal bulges, one on the side of the earth directly beneath the moon, the other on the side opposite. Think of those bulges as stationary, held in permanent position by the moon. As the earth makes its daily rotation the surface sweeps along beneath the bulging water, causing each point on every coast (with a few exceptions) to experience two high tides and two low tides during each tidal day.

Every two weeks, when the moon is in its full and new phases, the tidal influences are greatest, causing *spring tides*. The name has nothing to do with the seasons. It means only that the combined gravitational pull produced when the moon, sun, and earth are aligned increases the height of the oceans' bulges, causing the tides to "spring" to their highest levels.

When the moon is in its first and last quarters there will be *neap tides*, the lowest tides of each month. The sun's gravitational pull on the oceans, though weaker than the moon's, is still significant. When the sun and moon are at right angles to each other—during the moon's half-phase— the sun is creating bulges of its own on the sides of the earth away from the two moon bulges. The four bulges have the effect of partially canceling one another, muting the extremes and making high tide lower and low tide higher.

All this would be quite straightforward if the earth and the moon circled in perfect orbits and the entire surface of the earth were covered with an ocean of uniform depth. But there are complications. The highest tides of all, for instance, occur at least twice a year, when the in-phase alignments of sun, moon, and earth coincide with the *perigee*, or nearest point, of the moon's elliptical orbit of our planet. When the moon is nearest, it exerts about 20 percent more pull than when it is at its farthest point, or *apogee*.

The moon also goes through a cycle of *declination*, varying in its orbit so that some years it is north of the equatorial plane and some years south of it. The sun's declination (caused by the earth's position on its tilted axis) puts it south of the equator in the winter and north of it in the summer. As a result, the bulging tides tend to oscillate north and south as they sweep around the globe and in some places vary from the usual twice-daily or *semidiurnal* pattern discussed above. Places that are aligned with the moon when it is in declination are likely to experience a single, or *diurnal*, tide every 24 hours because the centrifugal bulge on the other side of the earth misses them.

Still other complications arise. As the bulge of tide sweeps across the oceans, friction on the bottom slows it, causing the bulge to lag slightly behind the moon. It is also reflected from the edges of continents, refracted as it passes through different ocean depths, diffracted when it passes through channels between landmasses, and enhanced or subdued by wind and barometric pressure. In large ocean basins the tide wave is so long that it is affected by the Coriolis force, causing it to turn to the right in a clockwise direction around the oceans in the Northern Hemisphere, and to the left in a counterclockwise rotation in the Southern Hemisphere. Variations of these kinds cause the tides in some coastal regions to be regularly irregular—that is, each day there will be a high-high tide and a low-high tide, as well as a low-low tide and a high-low tide. These are known as *semidiurnal mixed tides*.

The range between high tide and low tide varies greatly from place to place. Small islands surrounded by deep water in mid-ocean are washed by tides only about one foot in height. In contrast, much higher tides reach coasts preceded by shallow continental shelves. Bays and estuaries that funnel the incoming tide experience the greatest variations between high and low water. The differences are explained by bottom and shore configurations. As the width of an incoming tide wave is restricted by the sides of the bay, the wave's energy is concentrated toward the center and it increases in height. The effect is counteracted if the bay or estuary is very long—friction on the sides and bottom reduces the size of the wave and can erase it altogether.

The configuration of a coast can make all the difference. Tides on Nantucket Island rise only about a foot, while a few hundred miles north in the Bay of Fundy they rise over 40 feet. The Caribbean side of the Panama Canal has tides a foot high, while those of the Pacific side average 14 feet. Giant tides of 30 feet or more occur in the Bay of St. Malo on France's Brittany Coast, Turnigan Arm, near Anchorage, Alaska, and the Bay of Fundy—all narrow, funnel-shaped bays where incoming tide piles up and is shoved to the head of the bay.

The complex arrangement of forces and variables that influence the tides has made them among the most baffling of natural phenomena. Ancient thinkers in the Mediterranean region were at a disadvantage because the only tides they saw regularly were smaller than three feet high. Puzzlement increased after mariners dared to sail through the Gates of Hercules and challenge the Atlantic, where they were amazed to find tides 10 times higher than the Mediterranean's. Aristotle, Pytheas, Posidonius,

Pliny, and a few other Greeks and Romans recognized that lunar and solar influences were at work and guessed that the heavenly bodies must exert some form of attraction on the earth. Others were not so sure. Solinus, a third century B.C. geographer, wrote, "Our physicists assure us that the world is an animal which is formed of various members, that it is animated by breath and directed by intelligence. Now just as there is a vital breath in our own bodies, so the depths of the sea are in a way the nostrils of the world; when the world breathes in, it lowers the level of the seas; when it breathes out, the sea levels rise again."

The concept of a living, wheezing sea has found its way into legends and lore. Naturally, a creature as enormous as the oceans would have a powerful effect on humans and other things living near it. Farmers in some coastal regions still believe that seeds sown during an ebbing tide will not germinate. Such widely separated cultures as the ancient Greeks and Romans, western Europeans since the Middle Ages, Chilote Indians of southern Chile, and Haida Indians of the Queen Charlotte Islands in the Pacific Northwest have believed that humans die only during an ebbing tide. Shakespeare seems to have been familiar with the idea: He made Falstaff in *Henry IV* die "e'en at the turning o' the tide." In *David Copperfield*, Charles Dickens has his character Mr. Peggotty say, "People can't die, along the coasts, except when the tide's pretty nigh out. They can't be born, unless it's pretty nigh in ...."

As late as the seventeenth century scientists such as Galileo and Descartes were still perplexed by the mechanics of the tides. It wasn't until 1687, when Newton completed a detailed analysis of the lunar effect on the oceans, that science recognized what made the waters flood and ebb every day. Armed with Newton's Law of Universal Gravitation—the force of attraction between any two objects is directly proportional to the product of their masses and inversely proportional to the square of the distance between them—scientists finally understood why the oceans are tugged by the moon and the sun, and why they are tugged more urgently by the moon.

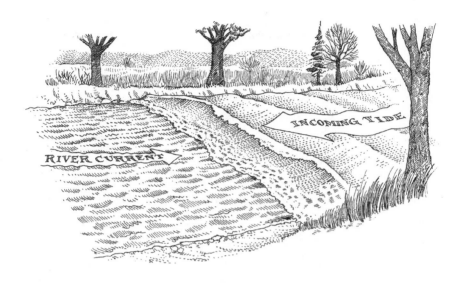

TIDAL BORES

These unusual solitary waves occur when an incoming tide is funneled into a long bay or estuary at the mouth of a river. As the front of the rising tide meets the gradually shoaling bottom of the river mouth it rises in height and forms a distinct and often quite large wave in conflict with the downflowing current of the river. As the tide continues to flood it rushes upstream in a wall of water with the level behind it higher than in front of it. As it progresses it gains energy from the funneling effect of the river as the river grows more narrow and shallow.

Most tidal bores are only a few inches high, but on certain rivers, especially during spring tides, they reach epic dimensions. The Amazon's bore can reach a height of 25 feet and surge as far as 300 miles upstream like a churning, mobile waterfall. It is large enough to be heard rumbling like thunder as far as 15 miles away and lasts so long that there are often three or four widely separated bores marching upriver at the same time. China's Chien-tang River has a tidal bore averaging 8 to 11 feet high that travels upriver at a rate of over 12 knots, creating a hazard to boat traffic in the port city of Hangzhou. In Bay of Fundy rivers like the Petitcodiac in New Brunswick and the Salmon in Nova Scotia, tidal bores are large enough to support a minor tourist industry: Passengers in inflatable rafts surf upstream on the bores, then, when the tide turns, ride down with the current.

Bores are known by various names. The Amazon's is *pororoca*, "the rock crusher." On England's River Severn it is the eagre, on the Seine in France, *mascaret*, on the Hooghly River in Calcutta, India, *bahu*. The bore at the mouth of the Colorado River used to be known as *el burro*, before it was made extinct by heavy siltation of the Colorado delta and the diminishment of the river's flow.

TIDAL RACES, WHIRLPOOLS, AND OVERFALLS

Tide is always accompanied by *tidal currents* traveling horizontally through the water. When the tide is rising, the current floods. When the tide falls, the current ebbs. It is slack, or free of current, only when the tide slows and reverses direction.

Tidal currents can be substantial, often reaching 4 knots at the mouth of San Francisco Bay and 10 knots in Seymour Narrows, Alaska. They can be recognized on the surface because they create choppy waves and turbulence where they encounter wind waves that force them to become steeper and break.

Where flowing and ebbing currents are forced through straits and channels they can reach such high velocity that they are said to race or rip. In places like the Queen Charlotte Straits, British Columbia, they race through narrows like rapids on a river, reaching speeds clocked as high as 18.4 miles per hour. They create a real hazard in the straits between the Orkney and Shetland Islands in the North Sea and around the Aleutian Islands, where waves as high as 15 feet are thrown up and occasionally sink small craft and sweep over the decks of larger ones. Like current in a river, a tidal race as it passes headlands, rocks, jetties, and other stationary objects will create eddies and whirlpools, some large enough to be dangerous. If the racing current passes over a reef or a shoal of shallow water at the mouth of an estuary, a waterfall of seawater known as an *overfall* can appear.

Any tidal race, whirlpool, or overfall can be dangerous, but few are equal to the reputation given them in mythology and literature. The most famous are found in the Strait of Messina between Italy and Sicily. On the Italian shore is a rock named for Scylla, the sea monster of Greek mythology with a woman's form and the heads of six ferocious dogs growing from her groin. In the *Odyssey*, the dogs of Scylla attack Odysseus's ship as it passes their cave and devours six members of the crew. A whirlpool on the Sicilian side of the strait is named Charybdis, for the goddess who stole Hercules's oxen and in punishment was cast into the strait and turned into a monster. She drinks huge quantities of water three times each day, swallowing anything floating on the surface. Odysseus slipped through the strait once without mishap, but during his second passage the whirlpool sucked his ship under and nearly drowned him.

The currents in the Strait of Messina run as fast as six miles per hour, creating dramatic boiling eddies and foaming streamers and an overfall where the tidal race passes over a shoal. Whirlpools often appear, but though they look menacing, they are a threat only to small boats.

The Maelstrom (Dutch for "whirling stream") is a whirlpool or series of whirlpools that form as strong tidal and coastal currents funnel between islands off the west coast of Norway. The power of the tidal current causes the whirlpools to become quite large, taking on the shape of an inverted bell, wide at the top and narrow at the bottom. They are largest when they first form and diminish in size as they are carried by current through the tideway. Fishing boats have reportedly been dragged into the whirlpools and people have drowned, but the hazard is not as terrible as Edgar Allan Poe described it in his story "A Descent into the Maelstrom":

> I perceived that what seamen term the chopping character of the
> ocean beneath us, was rapidly changing into a current which
> set to the eastward. Even while I gazed, this current acquired a
> monstrous velocity. Each moment added to its speed—to its headlong

impetuosity. In five minutes the whole sea as far as Vurrgh, was lashed into ungovernable fury; but it was between Moskoe and the coast that the main uproar held its sway. Here the vast bed of the waters seamed and scarred into a thousand conflicting channels, burst suddenly into frenzied convulsion heaving, boiling, hissing—gyrating in gigantic and innumerable vortices, and all whirling and plunging on to the eastward with a rapidity which water never elsewhere assumes, except in precipitous descents.

RIP TIDE

20

THE DYNAMIC BEACH

A few hours on the open sea is enough. The novelty wears off. The sky, the water, the rolling deck become monotonous, and you soon start wishing for a place where the waves end. It's been said before: The best part of the ocean is the beach.

In my childhood I spent a lot of time exploring the shore of Lake Michigan. To me calling the lake an "inland sea" did it an injustice. It was no mere sea, it was an ocean, almost an arm of the Atlantic, connected by canals and navigable rivers to all the oceans and seas. Its borders were invisible beyond unbroken horizons and it was sailed by ships flying the flags of dozens of nations. I imagined it must share its color, clarity, and tideless extremes of calm and violence with the Mediterranean.

The beaches along what is now Sleeping Bear Dunes National Lakeshore were long and uncrowded. Because many people still clung to a vestige of the old attitude that the shore was undesirable property—too cold, too windy, too plagued by blowing sand and biting insects—it was unusual to see anyone below the bluffs between Empire and Otter Creek or around the point west of Glen Haven. To a child few things are more wonderful than an empty beach on a July morning.

The wonders took many forms. In lines of debris marking the reach of the last storm I found drowned butterflies and ladybug beetles, plastic

SWASH MARKS

BACKWASH

SWASH

SAND DOME

LOW TIDE STEP

toys washed pale of color, lost fishing lures. The aluminum floats from the nets of commercial fishermen seemed so exotic they could have been lost by Polynesians. There was driftwood shaped like herons and gazelles. Silver polished tree trunks as large as Roman columns washed up in storms after drifting waterlogged through the lake for 50 or 60 years. In the brightly colored gravel at the water's edge were stones worn round as marbles and bits of polished glass the color and texture of candy half-dissolved on your tongue. There were gulls and terns soaring above the waves and pencil-legged sandpipers running on the wet sand.

Best of all, everything changed. Every day the beach was different. A strong west wind would bring waves that swept miles of shore clean or washed up the wrecked hull and ribs of an old lifeboat. Each spring, after a winter of grinding ice and pounding surf, the beach would be transformed. Trees that had stood on precarious banks would be gone, their banks stolen from beneath them. Where the beach had been wide, it would be narrow; where it had been narrow, it would be wide. Rocks the size of bushel baskets were gone and new ones cluttered places that had been rock-free the year before. There was novelty everywhere, every day.

It's in the nature of water and land to change one another. Spend enough time on the shore of a lake or ocean and the boundaries between water, land, and air begin to blur. The changes are even more striking along an ocean. The land shapes the ocean and the ocean is constantly shaping the land. Waves collapse on the beach, dislodging rocks, undermining cliffs, transporting sand. Even the less obvious changes have far-reaching impact. Every breaking wave creates millions of bubbles. As the bubbles burst they toss droplets of salt water into the air, where some of them are caught by the wind and carried aloft to become seed for rain. Water vapor requires a bit of matter to anchor to when it condenses into droplets, and those microscopic flecks of salt are fine for the job.

The margin between water and land is made of identifiable elements. A *coast* is a large feature of the landscape that can extend inland several miles along hundreds of miles of shore. A *shoreline* is the place where water and land meet. A *beach* is the strip of land subject to change by waves and tides.

A beach's character is determined in part by the kinds of stone and other materials that collect on it. In the British Isles, many beaches are composed of shingle—small, flattish stones eroded from cliffs of sedimentary rock. Beaches along parts of the Alaskan coast are made of round cobbles the size of melons. The grayish-green sands of the Oregon and Washington coasts are composed of basalt. Along much of the east coast of the United States, wide, gently sloping beaches are composed of fine yellow or whitish sand caused by the disintegration of granite to its primary constituents, feldspar and quartz. Some volcanic islands in the Pacific have lovely beaches of pure black sand produced from volcanic rock that has been ground to powder by water and wind. On the windward side of Tahiti, for instance, the black, volcanic beaches collect rock carried down from the interior by rivers; on the lee side of the island are white sands of ground coral.

All sandy beaches are evidence of the power of waves. Like the shavings that heap around the feet of a woodcarver, sand is a by-product of work. Thousands of years of bashing, crushing, and grinding go into it. A rocky promontory jutting into a sea or lake is attacked by waves. They approach it from several angles at once, concentrating their energies on it. First the rock is broken into large fragments by the impact of waves, especially by the pneumatic force of water traveling at high speed, slamming against cracks and holes in the rock, and compressing the air inside. The compressed air can cause the rock to explode, flinging fragments into the waves. Once in the water, the fragments are subjected to the constant action of waves carrying other fragments, like hammers and chisels, that shatter them into ever-smaller pieces. Suspended sand acts like fine-grade sandpaper to continue carving and polishing, taking the rock away one particle at a time. The water itself dissolves minerals from the rock, weakening it and causing it to crumble. Eventually it wears away entirely. Like current in a river always trying to smooth its channel, waves on a shore work constantly to straighten uneven coasts and smooth jagged shores.

Sand is the end result of all that work. Rocks in the surf tumble and roll in the ultimate rock polisher, their corners rubbed off and surfaces polished until they're smooth as porcelain. When waves are violent, you can hear stones knocking against each other with sudden, hollow, cracking sounds, like baseball bats against watermelons. Find a rocky beach with much banging and thumping going on, and you can be sure that down the shore, in the direction followed by prevailing winds and current, you will find a sandy beach.

Sand moves. That simple fact accounts for the constantly changing face of a beach. Every inrushing wave moves sand tirelessly, and thousands strike every beach every day. Figure it out: If an average wave has a period of 10 seconds, there are 6 per minute, 360 per hour, 8,640 in a day. Each one does its part to change the beach. If a grain of sand moves just an inch down the shore with every wave, it moves 720 feet in a day's time, nearly 50 miles in a year.

On some shorelines, the distance can be lineal, resulting in a steady migration of sand from a source such as a disintegrating stone headland, to a barrier island or offshore bar or scalloped beach tens or hundreds of miles down the shore. The movement of the sand along a coast is called *littoral*

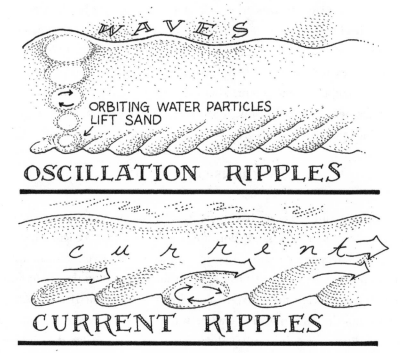

WAVES

ORBITING WATER PARTICLES
LIFT SAND

OSCILLATION RIPPLES

current

CURRENT RIPPLES

transport. Other beaches are closed systems, protected on each side by long headlands that halt the longshore currents that carry sand. In a closed system there is little net gain or loss of sand, but the sand is in constant motion anyway and the beaches still change.

The long, gently sloping beaches of summer are often replaced in winter by short, steep beaches. When winter storms pound a coast with large waves, tons of sand are carried from the beach and deposited on offshore bars parallel to the beach. In spring and summer, when the waves are smaller, they dismantle the bars and return the sand to the beach.

Many unhappy property owners have learned that it is natural for shorelines to retreat. The erosion of rocks and the transport of sand cause most ocean shores to recede by a few feet a year. Storms can eat land much faster, taking 50 or 100 feet in a single night, washing away houses and roads and scooping out bays where pastures and forests once grew.

In an effort to stop their retreating beaches, many people construct perpendicular breakwaters of rock, concrete, or timber called "groins." Sand always accumulates on whichever side of a groin, jetty, or breakwater faces in the direction of longshore drift. The problem is, as sand accumulates on the updrift side, it is reduced on the downdrift side. A property owner can solve the beach-erosion problem, but only by creating an erosion problem for the neighbor down-current. That neighbor must then build a groin, as must the next neighbor in line, one after another down the beach. Some shorelines are studded with groins for many miles.

PATTERNS ON A BEACH

Every feature of a beach has a story to tell about change and motion. The *ripple marks* that rib the sandy bottoms of lakes and oceans occur wherever there are waves and currents. They're symmetrical or asymmetrical, a few inches high or as much as 10 feet high, in shallows and on the ocean floor 18,000 feet beneath the surface. In shallow water, symmetrical *oscillation ripples* look like the mirror image of waves, as if they were left on the bottom like tracks. In a sense they *are* tracks, formed by the circular, oscillating patterns of water particles beneath the crest of each wave. As water particles orbit near bottom they lift particles of sand and rearrange them and gradually shape them into waves of ripples. The uniformity of the water's movement creates row after row of symmetry.

But ripple patterns are usually at least a little disorderly. *Current ripples* form when water runs in a single direction, as in rivers, creeks, floods, and longshore currents. They are asymmetrical, long and tapering on the side the current comes from, steep and short on the opposite side. Along beaches, currents running parallel to shore can work in conjunction with waves to form a combination of oscillation and current ripples.

Current ripples have a peculiar habit. If the current is flowing faster than about two feet per second, an eddy forms behind the steep, lee side of the ripple and transports sand toward the long side, causing the ripple to move slowly against the current. When the current flows faster than about 2.5 feet per second the ripples can't stand up against the pressure and are swept away altogether.

The final gasp of a wave is a thin sheet of foamy water flowing as high up the beach as momentum can take it. When this uprush, or swash, finally expends its energy and stops, some of the water sinks with a breathy sigh into the sand and the rest slips back down the slope of the beach as backwash. The swash is less than a quarter-inch deep at its upper edge—thin, yes, but thick enough to push a line of sand particles ahead of it. When the swash finally recedes, it leaves those grains of sand behind in a lazily undulating line called a swash mark. It remains until covered by a higher swash. On a falling tide, or when a train of large waves is followed by a train of smaller ones, a series of swash marks leave a record of the recent history of waves. But watch long enough—until the next incoming tide, perhaps—and you can see this record erased and a new one composed.

When a wave rushes farther than usual up a beach, it can reach sand that is dry enough to contain air between the particles. Dry sand allows quick infiltration by the swash. But as the water sinks, it displaces the air trapped among the sand particles, forcing it upward where it emerges in a series of bubbles. The bubbles form tiny, temporary openings in the sand known as *pinholes*. When a swash invades new territory, where the sand is very dry, water sinks only a few inches into the sand and forms a wet lid over a layer of dry sand beneath. The next uprushing swash sinks farther, beyond the saturated sand, and forces the air beneath to gather into pockets. Now, instead of escaping in many small bubbles, it escapes in a few large ones. Each of these large rising bubbles pushes sand above it until it emerges on the surface as a sand dome measuring a few inches across and a quarter-inch to a half-inch in height.

Backwashing water often leaves a series of diamond-shaped *backwash marks* where the flowing water has eroded the sand. The marks are oddly similar: diamond shaped, crisscrossed by tiny valleys about a quarter-inch deep. As the backwash continues to flow downslope it gains velocity and carries a great deal of sand with it in suspension. When this rushing mixture of water and sand reaches the level of the lake or sea, its force carries it a little farther, creating a foot-high or smaller curling wave known as a *backrush breaker*. The smaller particles of sand in this wave are held in suspension and carried back up the beach by the next incoming wave, but larger, heavier ones settle to the bottom and accumulate. This accumulation of sand and gravel, just below the level of the backrush breakers, builds into an abrupt and uniform *low-tide step*. It is usually about a foot high, with soft, yielding sand along the top of it and coarser sand, pebbles, and stones below it.

When a tide recedes, water left behind seeps gradually through the sand and makes its way downslope toward the receding shore. Eventually the water emerges on the surface and creates a pattern of drainage routes known as *rill marks*, each of which branches into smaller rill marks as it flows downslope, creating a network of branches similar to a river system's, only working in reverse.

A mysterious and common feature of many shorelines are *cusps*: crescent-shaped depressions arranged in regular order along the front of a beach. On Pacific beaches they tend to be 15 to 100 feet long, but elsewhere they vary greatly in size and shape, and have been found on beaches of all sorts, from sand to cobblestones, from open ocean beaches to sheltered coves.

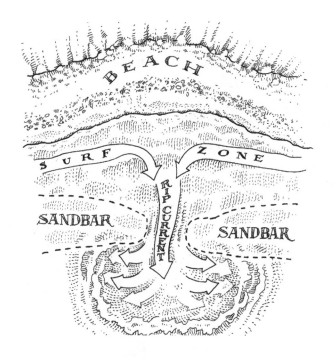

RIP CURRENTS

Diligent parents everywhere believe that when children cross their eyes they risk permanent damage, that swimming within half an hour of eating a meal results in cramps, and that the greatest danger of all is the undertow. When we were kids, my cousins, brother, and I would torment our mothers by crossing our eyes the entire time we waited for those interminable 30 minutes to pass. None of us believed for a moment that we would be afflicted with crossed eyes for the rest of our lives, and we were confident that we could swim to shore even while suffering the most terrible case of cramps. But the undertow was a different matter. The word itself conjured familiar fears. It suggested malicious intentions, an evil intelligence in the water. We were already convinced that dark powers lurked in the world and that if we put our guards down they would reach a clawed hand up, grab us by the ankles, and drag us beneath our beds. Besides, the physical existence of an undertow was plain to see. Every breaking wave climbed the beach in a surge of foamy water, paused, then retreated in a gravel-sucking rush to meet the next wave. That retreating water had to go somewhere. Any kid unfortunate enough to get caught in it would surely be pulled beneath the waves and never be seen again.

Lucky for us, the undertow as we imagined it does not exist. Retreating waves don't go far. They are engulfed by oncoming waves and tangled up in turbulence. No significant current flows seaward. Swimmers who lose their footing in breaking waves will be tossed about, maybe battered a little, but will end up about where they started. If they are swept anywhere, it will be a short distance down the length of the beach, carried along by longshore current. It's possible to get pinned momentarily by the weight of a wave breaking over you—and the weight can be considerable, and dangerous, if the wave is large enough—but that is not an undertow. The only dangerous undertows are found in rivers, particularly where current passes over low-head dams and plunges to create a circulating hydraulic current. On the beaches of oceans and lakes, undertows are not a threat. That monster can be put to rest.

But that is not to say there are no hazards in the surf. What many people mistake for an undertow is seaward-rushing water known as a *rip current*. This occurs when waves break continuously and in rapid succession on a submerged sandbar a short distance offshore. As the waves break they force water into the region between the sandbar and the shore, adding water faster than it can escape so that it builds up slightly higher than sea level. This mound of water flows laterally along the shore until it finds an escape route back to the sea across a low spot in the sandbar. As it flows over the bar, it dredges a narrow channel and builds into a continuous, fast-flowing current running straight out from the shore. But though a rip current can be powerful and fast, it is short-lived. After it passes beyond the zone of breaking waves it immediately disperses.

Rip currents are potentially hazardous, but they are not found on every beach and they are almost always visible from shore as an area of unusual turbulence. Swimmers caught in a rip can usually escape by swimming across it (they are rarely wider than 10 to 20 feet) or by riding it until it loses its force beyond the breakers. They can then swim parallel to shore a short distance and return with the waves.

On summer nights when I was a child, my parents sometimes built bonfires of driftwood on the beach and we sat up late cooking marshmallows, watching meteors, and listening to the waves. In the darkness, Lake Michigan seemed smaller, more lake than sea. Some nights it was so calm there were no waves at all, only a breathlike whispering cadence you sensed before you heard, and the stars were reflected in a mirror universe as far as we could see. Other nights the wind would be up and we would gather on the upwind side of the fire and watch sparks stream down the beach. Sometimes I noticed that when I put my ear to the ground on the high knolls above the beach I could hear each wave's thumping impact before the sound came through the air. I didn't know that sound travels 10 times faster through the ground than through the air. I thought I was hearing echoes or the thump of some other waves I had not yet discovered, and it didn't occur to me to ask my parents for an explanation. It was just another mystery. Until you reach a certain age everything is equally amazing. Waves can romp through the earth and sparks fly off to join stars and if you are quiet and listen carefully you can sometimes hear voices in the water.

ACKNOWLEDGMENTS

We're grateful to many friends and colleagues for reading portions of the manuscript, examining the illustrations, and offering their suggestions and support. Thanks especially to Gail Dennis, Kurt Schmidt, Debbie Behler, Jim Olson, and Doug and Anne Stanton. Thanks also to Robert Chapman; Terrie E. Taylor, D.O.; Emily Mitchell and Tim Nielsen of the Nielsen Design Group; Alexandra Smith; and Russell, Harry, and June Janis.

SELECTED BIBLIOGRAPHY

Allaby, Michael. 1992. *Water: Its Global Nature.* New York: Facts on File.

Bardach, John. 1964. *Downstream: A Natural History of the River.* New York: Harper and Row.

Barlow, Maude. 2008. *Blue Covenant: The Global Water Crisis and the Coming Battle for the Right to Water.* New York: New Press.

Bascom, Willard. 1980. *Waves and Beaches: The Dynamics of the Ocean Surface.* Garden City, NY: Anchor Press.

Berenbaum, May R. 1989. *Ninety-nine Gnats, Nits, and Nibblers.* Urbana and Chicago: University of Illinois Press.

Beston, Henry. 1992. *The Outermost House.* New York: Henry Holt and Company.

Briggs, John, and Peat, F. David. 1989. *Turbulent Mirror: An Illustrated Guide to Chaos Theory and the Science of Wholeness.* New York: Harper and Row.

Brittain, Robert. 1958. *Rivers, Man and Myths.* Garden City, NY: Doubleday and Company.

Burgis, Mary J., and Morris, Pat. 1987. *The Natural History of Lakes.* New York: Cambridge University Press.

Caduto, Michael J. 1990. *Pond and Brook: A Guide to Nature in Freshwater Environments.* Hanover, NH: University Press of New England.

Campbell, David G. 1992. *The Crystal Desert: Summers in Antarctica.* New York: Houghton Mifflin Company.

Cone, Joseph. 1991. *Fire Under the Sea.* New York: William Morrow.

Coote, Captain John O., ed. 1989. *The Norton Book of the Sea.* New York: W.W. Norton and Company.

Corbin, Alain. Translated by Jocelyn Phelps. 1994. *The Lure of the Sea: The Discovery of the Seaside in the Western World* 1750-1840. Berkeley and Los Angeles: University of California Press.

Crompton, John. 1988. *The Sea.* New York: Nick Lyons Books.

Croutier, Alev Lytle. 1992. *Taking the Waters: Spirit, Art, Sensuality.* New York: Abbeville Press.

Davis, Richard A., Jr. 1991. *Oceanography: An Introduction to the Marine Environment, Second Edition.* Dubuque, IA: Wm. C. Brown Publishers.

Darwin, Charles. 1962. *The Voyage of the Beagle.* Garden City, NY: Doubleday and Company.

Deming, H.G. 1975. *Water: The Fountain of Opportunity.* New York: Oxford University Press.

Dennis, Jerry. 2003. *The Living Great Lakes: Searching for the Heart of the Inland Seas.* New York: St. Martin's Press.

de Villiers, Marq. 2001. *Water: The Fate of Our Most Precious Resource.* Boston: Houghton Mifflin Company.

Dugan, Patrick, ed. 1993. *Wetlands in Danger.* New York: Oxford University Press.

Duxbury, Alyn C., and Duxbury, Alison B. 1994. *An Introduction to the World's Oceans.* Dubuque, Iowa: Wm. C. Brown Publishers.

Eblen, Ruth, and Eblen, William, editors. 1994. *The Encyclopedia of the Environment.* New York: Houghton Mifflin Company.

Einstein, Albert. 1934. *Essays in Science.* New York: Philosophical Library.

Fagan, Brian. 2011. *Elixir: A History of Water and Humankind.* New York: Bloomsbury.

Fishman, Charles. 2010. *The Big Thirst: The Secret Life and Turbulent Future of Water.* New York: Free Press.

Glennon, Robert. 2010. *Unquenchable: America's Water Crisis and What to Do About It.* Washington, DC: Island Press.

Halfpenny, James C., and Roy Douglas Ozanne. 1989. *Winter: An Ecological Handbook.* Boulder, Colorado: Johnson Publishing Company.

Hamblin, W. Kenneth. 1989. *The Earth's Dynamic Systems.* New York: Macmillan Publishing Company.

Hillel, Daniel J. 1991. *Out of the Earth: Civilization and the Life of the Soil.* New York: The Free Press.

Hynes, H.B.N. 1970. *The Ecology of Running Waters.* Liverpool: Liverpool University Press.

Kuenen, P.H. Translated by May Hollander. 1963. *Realms of Water: Some Aspects of Its Cycle in Nature.* New York: John Wiley and Sons.

Leopold, Luna B. 1974. *Water: A Primer.* San Francisco: W.H. Freeman and Company.

Lopez, Barry. 1986. *Arctic Dreams.* New York: Charles Scribner's Sons.

MacLeish, William H. 1990. *The Gulf Stream: Encounters with the Blue God.* London: Sphere Books.

Milne, Lorus and Margery. 1964. *Water and Life.* New York: Atheneum.

Moyle, Peter B. 1993. *Fish: An Enthusiast's Guide.* Los Angeles: University of California Press.

Muir, John. 1961. *The Mountains of California.* Garden City, NY: Doubleday and Company, Inc.

Newby, Eric. 1986. *A Book of Traveler's Tales.* New York: Viking.

Niering, William A. 1985. *Wetlands.* New York: Alfred A. Knopf, Inc.

Parker, Ronald B. 1984. *Inscrutable Earth: Explorations into the Science of Earth.* New York: Charles Scribner's Sons.

Pearce, Fred. 2006. *When The Rivers Run Dry: Water—The Defining Crisis of the Twenty-first Century.* Boston: Beacon.

Platt, Rutherford. 1971. *Water, the Wonder of Life.* Englewood Cliffs, NJ: Prentice-Hall, Inc.

Pollack, Henry. 2009. *A World Without Ice.* New York: Penguin.

Postel, Sandra. 1997. *Last Oasis: Facing Water Scarcity.* New York: W.W. Norton.

Pringle, Laurence. 1985. *Rivers and Lakes.* Alexandria, VA: Time-Life Books.

Prud'homme, Alex. 2011. *The Ripple Effect: The Fate of Freshwater in the 21st Century.* New York: Scribner.

Raban, Jonathon. 1993. *The Oxford Book of the Sea.* New York: Oxford University Press.

Rappoport, Angelo S. 1995. *The Sea: Myths and Legends.* London: Studio Editions, Ltd.

Reid, George K., and Wood, Richard D. 1976. *Ecology of Inland Waters and Estuaries.* New York: D. Van Nostrand Company.

Reisner, Marc. 1993. *Cadillac Desert: The American West and Its Disappearing Water.* New York: Penguin Books.

Richter, Brian, 2014. *Chasing Water: A Guide for Moving From Scarcity to Sustainability.* Washington, DC: Island Press.

Richter, Jean Paul, ed. 1970. *The Notebooks of Leonardo da Vinci (Volume II).* New York: Dover Publications.

Schumm, Stanley A., ed. 1972. *River Morphology.* Stroudsburg, PA: Dowden, Hutchinson, and Ross, Inc.

Sedlak, David. 2014. *Water 4.0: The Past, Present, and Future of the World's Most Precious Resource.* New Haven, CT: Yale University Press.

Short, Lester L. 1993. *The Lives of Birds: Birds of the World and Their Behavior.* New York: Henry Holt and Company.

Smith, C. Lavett. 1994. *Fish Watching: An Outdoor Guide to Freshwater Fishes.* Ithaca, NY: Cornell University Press.

Solomon, Steven. 2010. *Water: The Epic Struggle for Wealth, Power, and Civilization.* New York: HarperCollins.

Stevens, Peter S. 1974. *Patterns in Nature.* Boston: Little, Brown, and Company.

Teal, John, and Teal, Mildred. 1975. *The Sargasso Sea.* Boston: Little, Brown, and Company.

van Andel, Tjeerd. 1977. *Tales of an Old Ocean.* Stanford, CA: Stanford Alumni Association.

Watson, Lyall. 1988. *The Water Planet: A Celebration of the Wonder of Water.* New York: Crown Publishers, Inc.

Wendt, Herbert. Translated from the French by J.B.C. Grundy. 1969. *The Romance of Water.* New York: Hill and Wang.

INDEX

Abyssal floor, 204
Abyssal zone, 197
Adriatic Sea, 42
Aegean Sea, 185, 206
Age of Discovery, 186
Agriculture, 28, 29, 72, 156
Agulhas Current, 229
Ain ez Zarqa, 42
Aleutian Islands, 248
Aleutian Trench, 203
Algae, 29, 52, 58, 59, 65, 66, 120, 123, 146-48, 152, 155, 159, 161, 176-79, 181, 194-97
Alps, 46, 141, 144
Alps, Dinaric, 42
Alps, Japanese, 58
Amazon River, 72, 85, 125, 213, 247-48
American dipper, 95, 97, 99-100
American Society of Dowsers, 33
Amu Darya, 182
Anchor ice, 136
Andes, 95, 99, 120, 152, 188
Angel Falls, 11, 114-115
Angel, Jimmy, 114
Antarctica, 139-141, 160, 171, 198-199, 203
Antarctic Convergence, 198, 202
Antarctic Divergence, 198
Antarctic Ocean, 227
Antilles Current, 201
Appalachian Mountains, 90
Apparent color, 121
Apple bugs, 172
Aquifers, 2, 26, 29, 32, 33, 41, 45, 46, 51
Aral Sea, 182, 183
Arctic Ocean, 193, 227
Arroyos, 64, 90
Artesian springs, 41-42
Artesian wells, 41
Atlantic Ocean, 117
Atlantis, 235
Aughrabies Falls, 114
Australia, 29-30, 35, 92, 105, 155

Backrush breakers, 258
Backswimmers, 168
Backwash marks, 258
Baden-Baden, Germany, 53, 56
Badgastein Falls, 116
Baikal seal, 149
Ball ice, 129
Baltic Sea, 144, 213

Banff, Alberta, 53
Base-level streams, 63
Bath, England, 55
Baths, 2,19,40, 52-56, 58, 189, 190
Bay of Bengal, 232
Bay of Fundy, 214, 241, 245, 248
Beaches, 26, 126, 130, 134, 145, 155, 188, 191, 194, 211, 219, 223, 224, 225, 226, 237, 251-61
Beetles, 66, 91, 159-60, 166-67, 171, 197
Benthic zone, 197-98
Big Spring, 44
Black flies, 66, 104
Black Hills, 82
Black ice, 131
Blackmore, R.D., 62
Black Sea, 123
Blepharoceridae, 104
Blue duck, 99
Blue Spring, 44
Bogs, 61, 123, 156, 158
Boiling springs, 41
Bonneville Salt Flats, 174
Bores, 226, 247-48
Bouillens, Les, 45
Boundary layer, 103, 106
Braided channels, 84
Breakers, 191, 211, 222, 223-26, 228, 232, 258, 260
Bridalveil Falls, 115
Brine flies, 59, 177
Brine shrimp, 173, 176-80
Brooks, 61, 62

Caddisflies, 66, 92, 94, 103, 104, 152, 167
Canary Current, 201
Cape Cod, 226
Cape Horn, 227
Cape of Good Hope, 227, 236
Capillary action, 17, 27, 220
Capillary waves, 220, 228
Caracalla Spa, 56
Caribbean Sea, 119, 125, 178, 201, 237, 245
Caspian Sea, 186
Castalian Spring, 41
Castle bergs, 141
Catfishes, 105, 106, 155, 207
Cat's-paws, 220

Caves, 33, 35, 36, 42-44, 145, 167, 249
Cenotes, 41, 43
Challenger Deep, 203, 205
Chesapeake Bay, 213, 214
Chien-tang River, 247
Cichlids, 150, 152-53, 179
Cirque lakes, 145
Coasts, 29, 45, 141, 145, 189, 198, 199, 202, 212, 238, 245, 246, 254
Colorado River, 32, 85, 87, 125, 248
Columbia River, 174
Columbus, Christopher, 186-87, 200
Confined aquifer, 26, 42
Congo Basin, 158
Congo River, 72
Connate water, 36
Consumption of water, 15, 33
Convergence, 202
Coriolis force, 76, 199, 200, 203, 245
Crater Lake, 122, 146
Crayfish, 91, 147, 159
Creation myths, 14
Creeks, 1, 3, 5, 7, 11, 46, 61-68, 75, 88, 90, 123, 145, 171, 241, 251, 257
Crete, 235
Current River, 44
Currents, 3, 75, 103, 126, 138, 139, 141, 192, 193, 198, 199-202, 205, 214, 224, 225, 233, 248, 249, 256, 257, 259-60
Cusps, 258
Cyanobacteria, 123
Cyclones, 231

Darwin, Charles, 14, 87, 192
Dead Sea, 119, 174, 181-82
Death Valley, 145
Declination, 244
Deep circulation, 27
Deep-sea trenches, 203, 206
Delaware Water Gap, 82
Delphi, 41
Detritus, 66, 148, 150
Devil's Hole, 47
Devil's Hole pupfish, 47
Diatoms, 65, 125-26, 161, 197, 198
Diffraction, 225
Dinoflagellates, 126
Dippers, 99-102
Disappearing streams, 43
Discharge, 7, 85, 86
Distilled water, 18
Divergence, 198, 202
Divining rods, 33
Dobsonflies, 66
Dolphins, 126, 197, 208, 216, 232
Dowsing, 33-34
Dragonflies, 66, 166, 167

Earthquakes, 146, 233, 235, 239
Eastern Rift Valley, 177
East Grand Traverse Bay, 127
Eels, 154
Egeria, 40
Egypt, ancient, 20, 73, 74, 188
English Channel, 10
Ephemeral rivers, 90
Epilimnion, 148
Epsom salts, 55
Erosion, 28, 64, 76, 77, 88, 111, 145, 155, 205, 212, 256
Estuaries, 145, 158, 213-14, 229, 245
Euphrates River, 9, 16, 72, 74
Eutrophication, 123, 147
Evaporation, 5, 6, 7, 20-21, 27, 92, 130, 213
Everglades, 74, 158
Extreme Storm Waves (ESW), 228

Fairy shrimps, 179
Farming, 28, 30
Fertilizers, 2, 29, 123, 147, 153, 155, 183
Finger Lakes, 144
Fishing, 53, 63, 71, 78, 79, 147, 163, 182, 194, 199, 202, 249, 253
Flamingos, 177-79
Floeberg, 139
Floodplains, 156-58
Florida, 43-44, 46, 74, 146, 158, 201, 229
Foot ice, 130
Fontaine de Vaucluse, La, 44
Fountain Paintpot, 52
Fountains, 20, 22, 38-41, 43-44, 51, 52, 53, 56
Frazil, 136-37
Freshwater, 5, 21, 25, 40, 45, 61, 72, 137, 138, 140, 143, 149, 150, 156-58, 170, 176, 198, 211, 213, 215, 216, 233, 234
Freshwater marshes, 157-58
Frogs, 92, 159, 166, 172
Frost, 6, 127, 135-36, 145

Galápagos Islands, 46
Galileo, 246
Galveston, Texas, 232
Ganges River, 73
Garonne River, 44-45
Geothermal energy, 51-52
Geysers, 25, 49-52, 58, 177
Geysir, 50
Gibraltar Sill, 202
Glacier National Park, 145
Glaciers, 5-6, 25, 64, 75, 110, 111, 124, 133, 139-41, 144-45, 153, 161
Glaçon, 138-39
Gobies, 105

Goueil de Jouéou, 44
Grand Canyon, 85, 87, 97, 125
Grand Prismatic Spring, 52
Grand Traverse Bay, 127, 161
Gravity waves, 222, 225
Great Bear Lake, 144
Great Divine Order, 10
Great Lakes, 10, 143-44, 146, 162, 182
Great Plains, 32, 90
Great Salt Lake, 7, 59, 145, 173-74, 176-77, 179
Great Slave Lake, 144, 146
Greece, ancient, 19, 40-41, 45, 238
Greenland, 140-41, 203, 213, 227
Groins, 256
Grotto salamander, 36
Groundwater, 6, 8-9, 25-38, 41-42, 44, 50, 90, 94, 145, 146, 183
Growlers, 139
Guaíra, 112
Gulf of Argolis, 45
Gulf of Batabanó, 45
Gulf of La Spezia, 45
Gulf of Mexico, 201, 204, 214
Gulf Stream, 75, 119, 125, 199-201, 229
Guyots, 205
Gyndus River, 73
Gyres, 199

Hardanger Fjord, 116
Harlequin duck, 99
Hawaiian Islands, 239
Headwater streams, 62-63, 66
Hoarfrost, 135
Hooghly River, 248
Hot Springs, Arkansas, 53, 56
Hudson River, 72
Humpback whale, 207-08
Hurricanes, 141, 231-32
Hvítá River, 124
Hydrogen bond, 17-18, 85, 130
Hydrosphere, 5
Hydrotherapy, 58, 189
Hypolimnion, 148
Hyporheic zone, 93

Ice, 3, 5-6, 11, 13, 17, 25, 82, 90, 100, 111, 112, 123, 127-145, 150, 154, 159-60, 165, 169, 174, 181, 198, 213, 253
Icebergs, 139, 141
Ice field, 138-39
Ice floe, 138
Ice foot, 130, 138
Iceland, 49-52, 84, 124, 194, 226, 227
Ice sheets, 139-144
Indian Ocean, 123, 186, 193, 198, 204, 236, 239

Insects, 36, 40, 45, 59, 65-66, 95, 98, 99, 102-04, 146-49, 152, 154, 159, 166-70, 177, 197, 251
Intermittent streams, 90-92
Intertidal zone, 197
Irrigation, 16, 29, 32, 74, 79, 125, 182
Ishmael's Well, 41
Itchetucknee, 44

Japan, 19, 39, 51, 52, 58, 146, 187, 229, 238-39
Japan-Kuril Trench, 203
Java Trench, 204
Jigokus, 58
Jordan River, 176, 181

Karst regions, 43, 45
Kettles, 145
Kissimmee River, 74,
Kittatinny Ridge, 82
Koko Nor, 119
Krakatau, 235

Lake Athabasca, 144
Lake Baikal, 146, 149-50
Lake Bonneville, 174
Lake Chubb, 146
Lake Como, 144
Lake Constance, 144
Lake Erie, 3, 110, 112, 147, 162
Lake Geneva, 162
Lake Lucerne, 145
Lake Malawi, 152-53
Lake Mashu, 146
Lake Michigan, 3, 7, 10, 119, 121, 126, 145, 149, 161, 163, 219, 227, 251, 261
Lake Nabugabo, 153
Lake Nakuru, 177
Lake Natron, 177
Lake Okeechobee, 146, 158
Lake Ontario, 5
Lake Superior, 143-44, 149, 153, 182
Lake Tanganyika, 111, 146, 150, 152-53
Lake Texoma, 122
Lake Titicaca, 152
Lake Vanda, 174
Lake Victoria, 150, 153
Laminar flow, 82-83
Lampreys, 105
Larderello Hot Springs, 51
Larvae, 46, 59, 65-66, 92, 97, 102, 103-04, 106, 147-48, 152, 167, 176-78, 180
Leads, 139
Leopold, Aldo, 11, 72, 171
Leopold, Luna, 72, 75
Life, origins of, 14
Limestone, 26, 33, 35, 42-45, 52, 88, 145

Limnetic zone, 148
Littoral shelf, 155
Littoral zone, 147-48, 150
Loaches, 105
Loch Morar, 153-54
Loch Ness, 153-54
Loess, 123-24
Long Lake, 155
Low-tide step, 258
Lungfish, 92, 155

Madison Cave isopods, 36
Madison River, 53
Maelstrom, 249
Mammoth Cave, 33
Mammoth Hot Springs, 52
Manavgat River, 41
Mangroves, 156, 158
Marduk, 14
Mariana Trench, 203
Marshes, 64, 84, 88, 156-58
Mayflies, 66, 94, 103, 167
Maypu River, 87
Medicinal springs, 53
Mediterranean Sea, 10, 44, 125, 181, 185-
86, 202, 204, 206, 208, 213, 214, 245,
251
Menderes River, 72
Mesotrophic lakes, 147
Meteor Crater, 146
Middle Ages, 8, 50, 73, 185, 188, 246
Midges, 177
Mid-reach streams, 63
Minas Basin, 241
Mineral baths, 53-58
Mississippi River, 72, 74, 82, 85-87, 125,
201, 214
Missouri River, 144
Molecular structure of water, 16, 137
Mono Lake, 174
Monsters, 154, 188
Morning Glory Hot Pool, 52
Mosses, 161
Mount Fuji, 39
Mount Hood, 82
Mount Mazama, 146
Mount Parnassus, 40
Muir, John, 100, 102
Muska voda, 57
Myrtle Beach, South Carolina, 211, 217
Mythology, 13, 40, 73, 185, 249

Naiads, 40
Nantucket Island, 245
Narragansett Marine Laboratory, 207
Neap tides, 244
Needle ice, 135

Nekton, 197
Nepte oasis, 42
Neritic zone, 197
Neuston, 167, 171
Newfoundland, 144, 194, 199-200, 228,
238
Niagara Falls, 5, 110
Niger River, 125
Nilas, 138
Nile perch, 153
Nile River, 8, 9, 20, 63, 72, 73-74, 82,
111
North Sea, 248
Nymphs, insect, 102
Nymphs, mythological, 40

Oases, 42
Ocean currents, 8, 192, 199-201
Oceanic ridges, 205
Oceanic zone, 197
Ocean Sea, 185
Oceanus, 10, 14, 185
Ogallala Aquifer, 32
Ohio River, 120
Old Faithful, 51
Oligotrophic lakes, 147
Omul, 150
Orkney Islands, 248
Orontes River, 42
Oscillation ripples, 256
Osmosis, 214-15
Overfall, 248-49
Overturns, 148
Ozark Mountains, 44

Pacific Ocean, 97, 99, 123, 127, 186,
188, 191-93, 198, 199, 202, 203, 205,
213, 227, 229, 238, 245, 254, 258
Pack ice, 138-39
Palm Springs, California, 53
Panama Canal, 245
Pancake ice, 129, 138
Pangaea, 185
Paternoster lakes, 145
Pecos River, 90
Periodic Spring, 42
Permafrost, 136, 145
Persian Gulf, 186, 213
Peru-Chile Trench, 203
Photosynthesis, 15, 21, 65, 147, 148,
150, 159
Phytoplankton, 126, 146, 195, 197
Pinholes, 258
Plankton, 65, 122, 126, 147, 148, 152,
156, 197, 198
Plate tectonics, 36
Playfair's law, 81

Plunge pools, 111
Plunging breakers, 224
Pluvial lakes, 145
Pollution, 16, 29, 123, 147, 153
Polynyas, 138
Posidonius, 245
Precipitation, 5-10, 21, 27, 30, 90, 146
Profundal zone, 148, 150
Public baths, 52-53
Pupfishes, 46
Purification rituals, 19-20

Queen Charlotte Straits, British
Columbia, 246, 248

Radium Hot Springs, 53
Rainbow Springs, 44
Rapids, 5, 73, 77, 84, 86-87, 95-102,
105, 248
Red Sea, 123, 182, 213, 214, 235
Reelfoot Lake, 146
Reflected wave, 163, 225
Refraction, 225
Rhine River, 72, 82, 123
Ribbon Falls, 115
Rico-Cayman Trench, 204
Rift Valley, 150, 152, 153, 174, 177-78
Rift valleys, oceanic, 205
Rill marks, 258
Rime frost, 135
Rio Grande, 72, 82, 90
Rio Negro, 120, 125
Rio Puelo, 95, 120
Rip current, 259-60
Ripple marks, 256
River Severn, 248
Riverside Geyser, 51
River Styx, 8
Rivulets, 5, 28, 62, 75
Rocky Mountains, 1, 90, 145
Roman Empire, 52, 53
Rotifers, 161, 166, 177
Runoff, 7, 26, 28-29, 63, 64, 66, 90, 146,
147, 153, 174

Sacred springs, 39, 41
Sahara Desert, 7, 42, 51, 90
St. Lawrence River, 5, 112
Saline seeps, 30
Salmon, 66, 105, 124, 149, 154, 195, 216
Salt water, 32, 137, 198, 211-17
Salvadori duck, 99
San Francisco Bay, 248
Saratoga Springs, New York, 53, 56
Sardines, 150-52
Sargasso Sea, 125, 213
Scud, 46, 94

Scylla, 249
Sea ice, 137-39
Sea monkeys, 173-75
Seamounts, 205
Sea waves, 222-23, 227, 229
Secchi disk, 122
Sedimentary rock, 25, 36, 111, 254
Seepage springs, 41
Seiches, 161-63
Seine River, 10, 248
Seismic Sea Wave Warning System, 239
Semidiurnal mixed tides, 245
Seneca, 8
Sete Quedas, 112
Seymour Narrows, Alaska, 248
Sheet flow, 28
Shetland Islands, 248
Shorefast ice, 138
Shore flies, 59
Shorelines, 225-56, 258
Shrimps, 147, 150, 156, 161, 179, 207
Silver Springs, 44
Sinkholes, 32, 41-43, 45
Sirens, 208
Sleeping Bear Dunes National Lakeshore,
251
Sleet, 6, 135
Slovensky Raj, 43
Sludge, 137
Slush, 137
Snails, 46, 91, 92, 106, 147, 167
Snake River, 174
Snow, 1-8, 11, 25, 42, 51, 64, 65, 66, 69,
82, 90, 97, 115, 116, 123, 131-33, 135-
36, 139-41, 146, 159, 165, 231
Soborom Hot Springs, 51
Soda Lake fish, 177-79
SOFAR channel, 209
Soft water, 18
Soil, infiltration of, 27-28, 30, 146, 257
Solinus, 184, 246
Sorgue River, 44
Sound, underwater, 206-09
South Equatorial Current, 200
South Sandwich Trench, 204
Spas, 53-54, 190
Spilling breakers, 224
Spring tides, 243, 247
Staubbach Falls, 116
Steamboat Geyser, 51
Stoneflies, 46, 66, 94, 103, 154
Storm surges, 231-32
Straits of Florida, 229
Strait of Messina, 249
Strokkur, 50
Subduction zones, 36, 203
Submarine springs, 45

Submarine waves, 232-34
Sulis, 55
Surface layers, 197, 216
Surging breakers, 224
Sutherland, Donald, 115
Swamps, 92, 123, 156-58
Swash, 257-58
Swash mark, 257
Swell waves, 222
Syr Darya, 182

Tabular bergs, 141
Tardigrades, 94, 161, 166
Thames River, 72
Thaw lakes, 145
Thermal bar, 149
Thermocline, 148, 152, 199, 208-09, 214
Thermohaline circulation, 202
Tidal flats, 158, 241
Tidal race, 248-49
Tigris River, 9, 16, 72, 74
Torrent duck, 97-99, 102
Torrent midge, 104
Trains of waves, 163, 223, 228
Transparency of water, 122
Transpiration, 5, 7, 20-21
Trees, 29-30, 40, 63-68, 158
Trou du Toro, 44
Trout, 62, 63-64, 66, 68, 88, 97-99, 102, 105, 149, 154, 159
True color, 120-21, 123-24, 102, 105, 149, 154, 159
Tsunamis, 162-63, 231, 234-39, 241
Turbidity, 120
Turbulent flow, 82

Unconfined aquifer, 26
Undertow, 259-60

Valley of the Geysers, 51
Vapor, water, 5, 13, 36, 135, 216, 253
Velocity of a river, 83-84
Verglas, 136
Vichy, 54
Victoria Falls, 112, 114
Volcanoes, 49, 82, 111, 146, 203, 216, 235

Wadies, 90
Waimanugu Geyser, 52
Wakulla Springs, 44
Water boatman, 166, 177
Water scorpions, 166, 168, 190
Water spirits, 20
Water striders, 17, 167-71
Water table, 26-30, 32, 90, 157
Weddell Sea, 203

Wells, 8, 25, 29, 40-41
Western Rift Valley, 177
West Wind Drift, 199
Wetlands, 74, 156-58
Whales, 197, 198, 207-09, 216
Whirligig beetles, 152, 167, 171-72
Whirlpools, 87, 248-49
Whitecaps, 162, 222, 223
White ice, 131
Whitemarsh, R.P., 229
White Sands pupfish, 46
White Sea, 123
White water, 86, 98

Yellow River, 73, 123-24
Yellow Sea, 123, 124
Yellowstone National Park, 49, 51
Yosemite Falls, 115
Yosemite National Park, 100, 111, 115
Young ice, 138

Zemzem, 41
Zooplankton, 146, 153, 195, 197

Printed in the USA
CPSIA information can be obtained
at www.ICGtesting.com
LVHW012136101023
760766LV00028B/209